Introduction to EMC

Introduction to EMC

John S. Scott
Clinton van Zyl

Newnes
An imprint of Butterworth-Heinemann
Linacre House, Jordan Hill, Oxford OX2 8DP
A division of Reed Educational and Professional Publishing Ltd

A member of the Reed Elsevier plc group

OXFORD BOSTON JOHANNESBURG
MELBOURNE NEW DELHI SINGAPORE

First published 1997

British Library Cataloguing in Publication Data

A catalogue record for this book is available from the British Library

ISBN 0 7506 3101 5

Library of Congress Cataloguing in Publication Data

A catalogue record for this book is available from the Library of Congress

Typeset by Quorum Technical Services Ltd, Cheltenham
Printed and bound in Great Britain by Biddles Ltd., Guildford and King's Lynn

Contents

Dedication

This book is dedicated to the memory of John Scott. John died in May of this year, while busily planning his next book and putting the finishing touches to this one.

The way he went was more or less typical of the way he was: a man of tremendous intellectual energy, always deeply involved in his latest project to the exclusion of everything else. This was the spirit in which he began a long career dedicated to the clear and attractive presentation of technical information, from the 1958 *Penguin Dictionary of Building* and the 1965 *Penguin Dictionary of Civil Engineering* through two other technical dictionaries and other books to this, his latest work.

He is remembered by his family and friends for his kindness and inspiration, as much as for his desire to make people aware of the implications of technology on our health and the environment.

Clinton van Zyl, London, June 1997

Chapter 1
What is EMC ?

Electromagnetic compatibility (EMC) aims to prevent interference that disturbs radio or TV reception. The European Union's law (directive 89/336/EEC) on EMC came into force after four years' postponement on New Year's Day 1996. Under this law any European manufacturer or importer of electrical equipment must declare that his or her product will not cause interference in electronic or electrical devices and that it is itself immune from interference.

Achieving this is not easy but failure to attain EMC makes the importer or producer liable to criminal prosecution and to the resulting severe fines or imprisonment. Although especially important for radio and TV, the law is not restricted to them but concerns any device which might suffer or create interference. Even computers have suffered interference from nearby powerful transmitters and have themselves caused interference. Because this law has been called the 'most comprehensive, complex and contentious directive ever to emanate from Brussels', this book is needed to explain it in language understandable by people who are neither electrical nor electronic specialists.

EMC means avoiding disturbance to other devices by the creation of EMI (electromagnetic interference), whether through conduction, radiation or induction. Every unit for sale in the EU (European Union) must be so made that it neither creates nor suffers EMI. It must be immune from EMI (not susceptible). One type of EMI is radio frequency interference (RFI), at the frequencies of radio waves. Electrical noise is relatively feeble EMI.

An underlying reason for the EMC law is the Single European Act, in force since 1987, which compels the European Union, its

Commission, Council and Parliament to create an area without internal frontiers, and with free movement of goods, people, services and capital. The EMC law will enable electrical goods to circulate freely because of widespread confidence in their quality and safety.

Appendix B lists some appliances that suffer or create EMI but practically every electrical or electronic product can emit or be a victim of EMI unless the product is too tiny to noticeably affect anything else. EMI emanates in use from all switches, wires, cables, metal bars, electric blankets, battery chargers, fluorescent lamps, electric fences, badly fixed fuses, generating or welding sets, vacuum cleaners, videocassette recorders, and much else.

Tolerable interference

A level of interference that is low enough to be tolerable may be achieved sometimes merely with distance, by ensuring that the victim device is far enough away from the source of interference. Another help is to point the device in a direction which receives the minimum of unwanted radiation. Any piece of wire or metal pipe for cooling or heating can, if suitably oriented, act as a radio antenna to pick up unwanted noise and sometimes even re-transmit it.

The International Federation of Airline Pilots reports about 20 aircraft incidents annually caused by passengers who use mobile phones, CD players or other electronic devices on aircraft. The use of these appliances is forbidden because they can emit stray radio frequency (RF) waves of the same frequency as the beacons on the ground that help the pilot to stay on course. Even the remote control device for opening a garage door can create dangerous interference.

Most people are aware of interference that distorts radio or TV sound or pictures because of a neighbour's unshielded electric drill. This is unpleasant, although usually not dangerous, but interference can be disastrous, especially on a warship where space is limited, and several communication systems with their antennas compete, including powerful radar pulses that sweep the 360° horizon. Aircraft and cars are similarly vulnerable.

Dangerous interference

During the Vietnam war, on the flight deck of USS *Forestall*, an aircraft carrier, the release circuit of a rocket on one of the aircraft was unintentionally ignited by the ship's radar. The rocket was despatched, destroyed 27 stationary aircraft, and killed 134 people. Some 20 years later during the Falklands war, HMS *Sheffield* was sunk by a sea-skimming Exocet missile that could have been located by a radar system and then destroyed. But this radar had been switched off because it would have interfered with the sending of an important radio message.

Effective EMC could have prevented such disasters. Most EMC work until the early 1980s was aimed at the safety of military hardware. But by the 1980s it had become possible for electronics designers to show that the 5% of the total cost which covered the inclusion of EMC in the design was always highly profitable in preventing very expensive rethinking, often including costly stoppage of production.

Computer simulation

Using a computer to visualize a piece of equipment is neither cheap nor quick but it is much faster, less expensive and more useful than taking the completed unit to a laboratory and then having to await the test report. Computer simulation also allows the EMC designer to investigate the device, showing him or her where currents and fields are largest, which is almost impossible and in any case very slow with physical testing. Computer simulation thus enables corrections to the design to be introduced early at low cost.

Computer-aided engineering (CAE) programs and other 'tools' (computer methods) can help in the design of printed-circuit boards (PCBs) by showing where components and tracks on PCBs need to be placed differently. CAE of course precedes construction, identifying EMC problems to be solved by early re-design (Fig. 1.1).

Many other examples of interference exist. On a stretch of German motorway beside a powerful broadcasting station, one make of car with electronic controls would regularly stall. Eventually the motorway had to be screened from the broadcasting station by a wire mesh wall to solve the problem. Similarly, near the Brookman's Park radio station in north London, new push-button telephones continually suffered from the intrusion of BBC radio programmes until the interference was dealt with.

Mobile phone problems and polling

Mobile phones are held extremely close to the head. Doctors therefore wonder whether these phones can harm any of its very sensitive organs, such as eyes, ears or even brain. Advice given so far in the UK is to keep the antenna stretched out as far as possible to reduce the transceiver's power output. Corresponding procedure in Germany is mentioned at the end of Chapter 13.

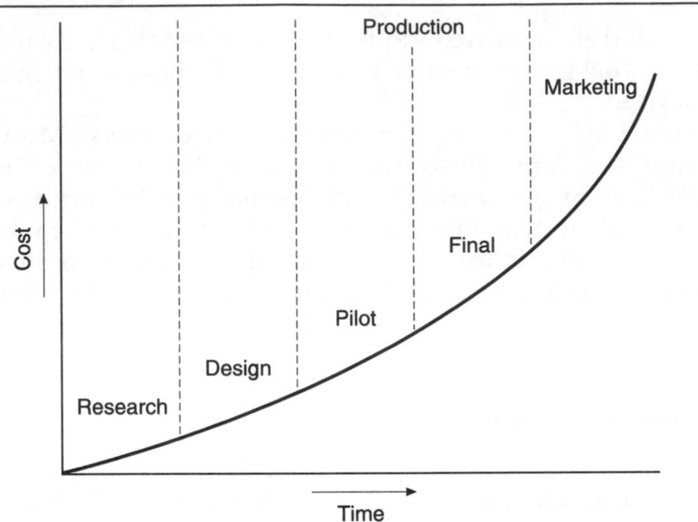

Fig. 1.1 Relative costs of introducing full EMC into a product at different stages between initial research and final marketing

Mobile phones have seriously interfered with the electronics of the car in which they receive a call, by activating the anti-skid system, or inflating an airbag. They have even varied the readings of a fuel pump meter in a filling station. In 1995 hospitals complained about interference with their equipment from the GSM mobile phones introduced to the UK from 1993.

Polling is another cause of interference which can be avoided if owners take care to switch off their mobile phones when they leave them unattended in a car. Every cellular base station, in its normal routine search for a subscriber, broadcasts an inquiry (called polling) which interrogates the required mobile phone if it is within reach. The resulting transmission has been known to activate a car alarm, apart from causing serious radio interference.

Electromagnetic waves

Electromagnetic (EM) interference depends on the relationships between electricity and magnetism and they need a little explaining. No hint of any connection between them was found before 1820. Nearly a century later the genius Marconi (1874–1937) felt, with most of his friends, that he had to presuppose an ether which carried EM waves because he could not imagine them travelling without a supporting medium. The ether is now forgotten and the word does not appear even in the all-embracing *Penguin Dictionary of Electronics.*

Whether this means that people are more intelligent or more stupid than in the early 1900s, I do not know, but a book describing EMC to people who are not electrical specialists must absolutely have a few notions about that fascinating, closely-connected pair, electricity and magnetism. So here is a short explanation of them.

Unlike speech, music and other sounds which are carried only by matter – usually by air but also by solids or liquids – EM waves need no medium to carry them. This caused confusion in the early days of radio, then called wireless. But the absence of any medium carrying EM waves has been demonstrated by the radio messages received across many hundreds of thousands of miles from the planetary rockets launched by the USA to explore the universe, not to mention radiation from the sun and background radiation from distant space.

Perpetual motion?

Few people can claim that they understand electric and magnetic waves perfectly. Perhaps the most surprising thing is that the two can be generated by each other. The flow of electricity generates magnetism but a moving magnetic field can also generate electric current. It seems like perpetual motion, which we have been taught is impossible. But in electromagnetism both the electric and the magnetic fields are moving.

Units of frequency

Table 1.1 identifies units and Table 1.2 shows that EM waves start below the different frequencies of radio, and end above them. Their rising frequencies with ever-shortening wavelengths continue up to infrared (heat waves), then to visible light, and the very penetrating X-rays and even higher.

For the reader's convenience the relationship between wavelength and frequency is summarized in Table 1.2, although it is described in Chapter 5.

Table 1.1 Names and sizes of units

Name	Cycles per second (hertz)
One hertz (Hz)	1
One kilohertz (kHz)	1000
One megahertz (MHz)	1 million
One gigahertz (GHz)	1000 million
One terahertz (THz)	1 million million

Table 1.2 Electromagnetic waves: approximate frequencies and wavelengths (many categories overlap)

Class of frequency	Frequency	Wavelength	Usual description
Audio or Voice	50 or 60 Hz	6000–5000 km	Mains power
	30–300 Hz	10 000 km–1000 km	Extremely low frequency, ELF
	300 Hz–3 kHz	1000 km–10 km	Very low frequency, VLF
RF (Radio)	3–30 kHz	100 km–10 km)	Low frequency, LF
	30–300 kHz	10 km–1 km)	(long wave)
	300–3000 kHz	1 km–100 m	Medium frequency, MF (medium wave)
	3–30 MHz	100 m–10 m	High frequency, HF (short wave)
	30–300 MHz	10 m–1 m	Very high frequency, VHF
Microwave	300–3000 MHz	1 m–10 cm	Ultra high frequency, UHF
	3–30 GHz	10 cm–1 cm	Super high frequency, SHF
	30 GHz–300 GHz	10 mm–1 mm	Millimetric
	300 THz	0.001 mm (1 μm)	Infrared (heat)
	1000 THz	0.0003 mm (0.3 μm)	Visible light
	10 000 THz	0.03 μm	Ultraviolet rays

Cosmic rays, X-rays and gamma rays, at even higher frequencies, are of less interest to EMC so are omitted.

A little history

EM waves are called Hertzian waves in mainland Europe because they were discovered by Heinrich Hertz (1857–94) in 1888. In that year he built an electric circuit in which a spark jumped across a gap. A few feet away was a copper ring complete except for a break between the two ends of the copper. When Hertz produced a spark in his electrical circuit, another spark jumped across the break in the copper ring. Spark transmission enabled Marconi, using Morse code, to send messages across the Atlantic from Cornwall to Newfoundland a few years later. Spark transmitters were superseded after 1904 when Ambrose Fleming (1849–1945) and Lee de Forest (1873–1961) developed thermionic valves.

The word electricity came from the Greek word 'elektron', which means amber. The ancient Greeks discovered that a piece of amber, if rubbed on cloth, will pick up small objects such as paper fragments. Rubbing a cloth on amber or another insulator such as plastic gives the insulator an electric charge, which enables it to pick up paper fragments (Chapter 6, Electrostatic discharge).

As early as 800 BC magnets were known in Greece, where lodestone was mined at Magnesia, in Thessaly. Seafarers used them to indicate the north direction, but no link between electricity and magnetism was demonstrated until 1820 when the great Danish professor Hans Christian Oersted (1747–1851) showed their relationship. Michael Faraday (1791–1867) confirmed Oersted's results on electromagnetic induction but also invented the transformer and much else.

Electricity and magnetism

Interference often happens only because people forget the connection between electricity and magnetism, that every electric current has its magnetic field and every voltage has its electric field. The magnetic fields of EM waves are measured in amps per metre (A/m) and their electric fields in volts per metre (V/m), although most often for EMI the units to be measured are in millionths, i.e. microamps or microvolts or even smaller. Having established that these fields exist it should be stressed that they only move as an electromagnetic wave if the current and voltage are changing (Fig. 1.2).

If the current and voltage alter, often merely by switching on or off, an electromagnetic wave with its two fields sets off on its space journey, possibly creating annoying interference in radio and TV sets on the way. Electromagnetic waves often start from an alternating electric current (AC) oscillating at radio frequency. The two fields are perpendicular to each other and to their direction of travel. Figure 1.3 shows that they vary uniformly, that is, they are sine waves. The current and voltage may be varying at one of the many frequencies of radio, or at frequencies below or above radio frequency levels. With rising frequency the variations of voltage and current become quicker, increasing the effects of induction, introduced in the following paragraphs. More information on

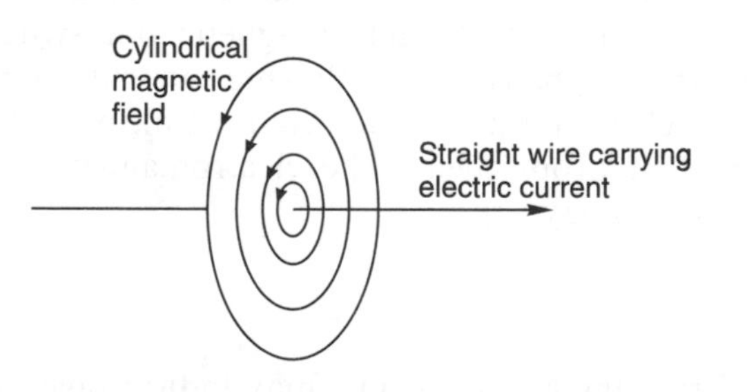

Fig. 1.2 Magnetic field produced by a current flowing through a straight wire

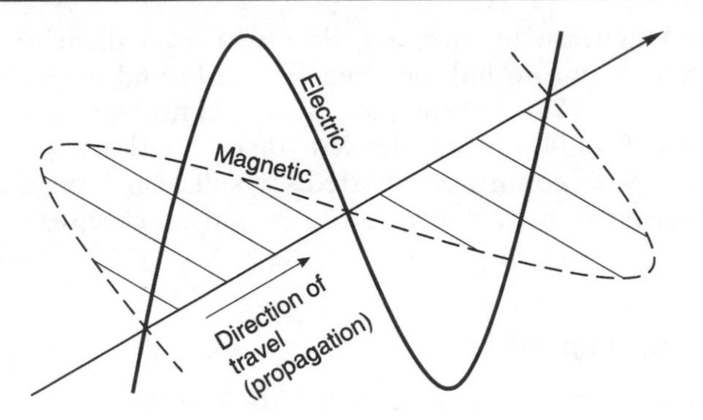

Fig. 1.3 Mutually perpendicular electric and magnetic waves

electromagnetic waves, their near fields and far fields is given in Chapter 5, Antennas.

So far we have considered current flowing in a straight wire. When the wire is not straight but coiled there are interesting complications without which electric motors or generators or indeed radio could not exist. A current passing through a wire coil produces, because of the coiling, a much stronger magnetic field around itself than through a straight wire. It is so strong that it can be looked upon as a magnet with poles, one magnetically north-seeking and the other south-seeking. To strengthen the electromagnet even further it may have a core of soft iron. Coils have relatively little resistance to direct current (DC) but to AC they form a considerable obstacle, called reactance or impedance. Any coil in an AC circuit is an inductor, sometimes called a choke. In a DC circuit it may not work as such because, although changing or pulsing DC is inductive, constant direct currents have no induction effect.

One type of iron-cored coil, often seen working suspended from a crane hook in a scrap yard, is called an electromagnet. A current flowing through it makes it strongly magnetic but switching off the current cuts off the magnetism. The flowing current enables it to lift tons of iron or steel from a pile of assorted metal, but when the crane driver switches off it drops them. The electromagnet is said to have been invented by Faraday.

Inductance

A person in a fury, raging at someone, may induce rage in him or her. Electrical induction, which all coils have, has some similarities with human rage, including two essentials, namely energy and nearness.

Every magnetic field generated by the current through the wires of a coil is reluctant to change and this is the idea of self-induction or self-inductance. When a current through a coil changes, another current opposing the change is induced in it. It slows down the rate of change of current and its accompanying opposing voltage is called a back EMF (electromotive force).

If another circuit (whether coil or straight wire) is close to a coil containing a changing current, the changing magnetic field of the first coil will induce a current in the second circuit (Fig. 1.4). This so-called mutual induction allows the design of electrical transformers and motors, but it also creates electrical interference and is interesting for other reasons. The symbol for an inductor used in electrical circuit diagrams is shown in Chapter 11, Filters, Fig. 11.2.

Electrical generation

When a conducting wire moves across a magnetic field it generates a current in itself by this movement. The amount and direction of the current are explained in textbooks of elementary electricity, which would occupy too much of this short book. If the conductor is stationary but the magnetic field moves, the result is similar. A current is induced in the wire. This is the basis of electric motors, because the magnetic field created by the current pushes against the original magnetic field and turns the motor round.

Though many shapes of wave exist, a common one, probably easy to visualize, is the sine wave (sometimes called sinusoidal) shown in Fig. 1.3. It is rare for such a 'pure' sine wave to exist because in the real world there are always other little bits of wave added to it, including some wavelengths of electronic noise, all of which help to distort it. But if it is assumed to be a pure sine wave the calculations are simpler because the mathematics of the pure sine wave are well known. The usual alternating current supply is

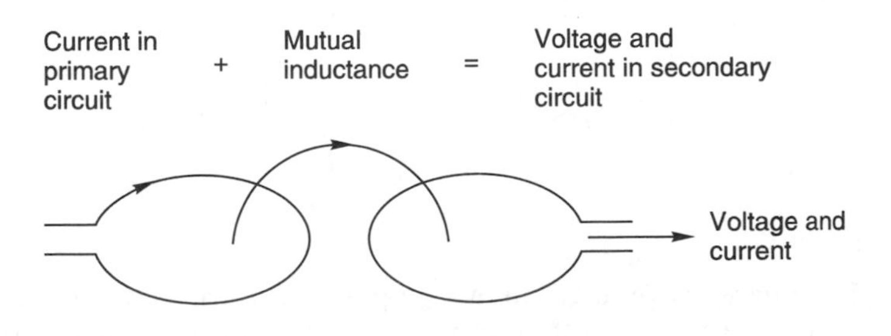

Fig. 1.4 Mutual inductance (after P. J. Fish, 1993)

generally taken to be a sine wave. In computers, square waves are exceedingly common, but a truly square wave, if it existed, would emit too much EMI to be desirable, so a 'rounded' square wave is preferred as in Fig. 1.5.

Power and dangers of microwaves

Powerful electromagnetic devices can be dangerous. For example, microwave ovens use considerable power, some 700 watts (W), enough to give a person a serious burn. They use the microwave frequency of around 2450 MHz. It is probably less well known that the harmonics from microwave ovens can upset satellite television.

Warships also use microwaves for their powerful radar pulses and it is possible for naval people to suffer a severe burn on the hand by touching a piece of pipe or other metal that has served as an inadvertent antenna receiving the ship's radar transmission.

But in most radio or TV sets the electromagnetic power received is extremely small because of their distance from the transmitter. It is much less than a millionth of the power used by a microwave oven. For this reason the power received by the antenna needs to be strengthened by an amplifier.

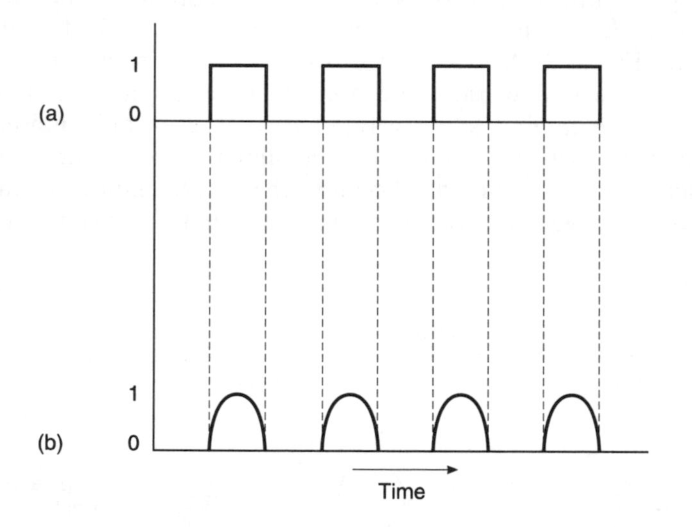

Fig. 1.5 Square wave used in a computer: (a) truly square, creating interference; (b) rounded with longer rise and fall times, consequently with less interference

Signal-to-noise ratio

Especially for satellite reception, the signal at the receiver is very faint, so the receiving antenna is fitted with a low-noise amplifier to make the messages understandable. A signal will not be intelligible if its accompanying electrical noise is equally strong. So the quality of a telephone or radio or other transmission is measured by a number called its signal-to-noise ratio (SNR). This is the strength of the signal plus noise divided by the strength of the noise. The higher the figure, the better is the quality of reception.

Common natural interference

Lightning seriously interferes with power cables, telephone cables and aircraft and is the most powerful source of electronic interference from natural causes. It comes from electrically charged clouds. The first strike, extremely strong, called the primary strike, often exceeds 100 000 A (amps). It is followed by 'return strokes' of around 10 000 A which may end after about one second. The primary strike has a rise time of about 1 μs (a microsecond, one millionth of a second). These disturbances, even if distant, are a permanent source of noise since some 2000 thunderstorms are taking place all the time in different places. Distant they are merely noise, close they can be dangerous. French research has triggered lightning artificially by shooting a 2.5 kg rocket into a likely cloud. The rocket, while climbing, unreels a 0.2 mm diameter wire which melts at a current of about 40 A.

Lightning conductors reduce the danger to a building from lightning. But recent research has shown that only about half the huge current of a lightning flash – some 50 000 A – reaches the earth through the lightning conductor. The other half passes through water or gas pipes or earths for protecting telephones or power supplies.

Another permanent natural magnetic event is the earth's magnetic field, with a strength from 19–54 A/m. This can be reduced to a thousandth of its normal level by the electric field accompanying a magnetic storm, often lasting several days. During a quiet period the earth's electric field is tiny, around 20 mV/km (millivolts per km) and much less in the oceans. But magnetic storms can multiply it to some 10 V/m, some 500 000 times as much. Transatlantic cables can thus develop an interference potential of 1000 volts (V) or more.

A NEMP (nuclear electromagnetic pulse) is fortunately not a common event but NEMPs have arisen from scientific investigations. They have so much destructive energy that they need a short mention at least. At 100 km distance a NEMP has a powerful electric field strength of 50 kV/m reached in a rise time of 5 to 10 ns (nanoseconds). It has a half-power duration of 200 ns. Consequently a long cable for power or telecommunications above ground may acquire a current pulse of several thousand amps. During the high-altitude tests of the 1960s in the Pacific Ocean, circuit breakers on power lines and burglar alarms went off in Honolulu, 800 km away, like popcorn.

A high-altitude rocket creating a nuclear explosion over the North Sea could paralyse any civil and military activities including mains power and telecommunications. The worst damage is likely against the most modern equipment but the EMC law requires no protection against a NEMP.

Stopping a car with an EMP

Because too many people are killed or injured in police chases of criminals, police in the USA are considering the use of an EMP (electromagnetic pulse) to stop a criminal's getaway car. Of several possibilities, the most hopeful is an electronic strip laid across the road ahead of a criminal's car.

Modern electronically-controlled cars are highly vulnerable to an EMP. As soon as the car-stopping strip is primed, the next vehicle to cross it triggers a discharge of some 100 000 V. A heavy current passes harmlessly through the bodywork and back to earth, radiating a pulse that demolishes the electronics without harm to the driver. An electric field to destroy a car's electronics is some 15 000 V/m, which does not harm people. Criminals needing a car that is not vulnerable in this way must find one with a 50-year-old engine, transmission and steering.

Another method under consideration is to miniaturize the electronics, creating a pulse simple and easy enough to be operated by a policeman alone in a car. The main objections are the enormous expense of fitting it into all police cars, as well as the danger that an EMP directed against a criminal's car could seriously damage other cars belonging to harmless civilians, which would be expensive to repair (*New Scientist*, 9 Nov. 1996).

Spectrum management

There are very many sorts of noise from natural causes but these are usually negligible nowadays compared with man-made interference. Exceptions should be made to this statement for satellite reception and radar reception, where natural noise is comparable in energy level with the signal. Radar pulses go out in megawatts but what comes into the radar receiver is in picowatts, a million million millionth of the outward pulse!

For many years it has been necessary for each radio transmitter to obey the IFRB (International Frequency Registration Board, a part of the International Telecommunications Union) and to broadcast only on its allotted frequency.

Broadcasting on another frequency is forbidden and regarded as environmental pollution. 'Spectrum management' is thus concerned with the spectrum of radio frequencies. Among existing natural resources the radio spectrum is virtually the only one which is indestructible, perhaps because it has no existence in material form. But it is suffering rapid, even accelerating incursions because of the growth not only of population, but also of the number of sources of man-made interference. Effective EMC helps in spectrum management by reducing interference.

One indication of the expansion of man-made interference is the amount by which radio organizations have been forced to increase the power of their transmitters. Thc Polish long-wave transmitter in Warsaw could cover all Poland and much of Europe in 1927 with only 12 kilowatts (kW) of broadcasting power. In 1974 this small unit had to be replaced by a 2000 kW transmitter, more than a hundred times more powerful.

Any changing electric current radiates (broadcasts) electromagnetic waves, as explained earlier. The higher the frequency of change, the faster is the rate of change of current, and the more effective the radiation. Near a switched-on home computer a portable radio receiver can suffer severe interference from the stray electromagnetic waves sent out by the rapidly-changing currents in the computer. Usually if the radio set is moved a yard away from the computer the interference ceases to be noticeable because most home computers emit very little energy. Emissions from some computers have, however, been picked up on a (VHF) TV receiver and the computer's display was seen on the TV screen.

Several megahertz is HF (high frequency). The higher the frequency or rate of change of current, the higher the likelihood of induction and transfer of interference to another device. But computers are not designed to broadcast so, fortunately, they do not do it very well. There are millions of other objects containing microprocessors that can unintentionally create interference because of their rapidly, changing electric currents. Low-power devices, such as portable electric games, hand-held computers and remote controls, emit HF energy that can be picked up on the medium- and short-wave bands.

When we rashly put a portable radio near a working computer, we notice that the orientation of the radio antenna matters. There is a best orientation for radio reception and a worst orientation, normally 90° different. All these factors – distance, orientation and rate of change of current – are important in eliminating or reducing electronic noise.

Troubles with very-large-scale integration

Miniaturization in modern electronics, including very-large-scale integration (VLSI), has created tiny electronic circuits, that are used in interplanetary rockets as well as in computers and many devices for automatic control. Circuits are thus closer together than was possible formerly, making interference hard to avoid unless it is tackled in the design stage of every electronic product. Careful design is essential so as to avoid mutual interference between closely-spaced circuits.

But with VLSI it is not its minuteness alone that makes interference more likely. Electronic devices are continually speeding up, and faster devices cause more interference. In 1989 it was estimated that microelectronic devices were either halving in size or doubling in capacity every two years. Where the product designer has the freedom to install slower logic in preference to fast logic, there will be less interference (Fig. 1.5).

The seriousness of the problem is evident in view of the vulnerability of transistors and switching diodes. These can be completely destroyed by a pulse of energy lasting only 1 μs and totalling at most 0.1 millijoule (mJ). One-hundredth of this amount causes temporary malfunction. Such a pulse and much more can be provided from a person's finger carelessly touching a contact point.

Computers are convenient but they involve fast automatic power-supply switching which is itself getting faster. The many advantages of automatic switching are counterbalanced by the radio disturbances they cause. Frequencies emitted at switching are from 10–50 kHz, but power transistors have reached 1 MHz. This is not altogether a disadvantage because as the wavelength shortens with increasing frequency the dimensions of filters as well as of the screens around them can be reduced. But power transistors, switching hundreds of volts and amps, are not like the infinitesimal switches in computers dealing with as little as 5 V and tiny currents.

Conducted interference

Interference can also be caused not directly by radio waves travelling through space but by electrical conduction into a victim appliance along a power cable or other wiring. Any metal, especially a cable for power or telephone or other purposes, can act as an antenna to pick up unwanted radio energy and conduct it into a device where it will cause disturbance. Other sources of noise entering power cables include switches whether in substations, industry or homes. Any switching on or off disturbs the power supply and the disturbance is led along the cables into other consumers' supplies.

The 'crosstalk' known to every telephone user also describes transfer of unwanted power or interference from one circuit to another in many systems remote from telephones. The advice of Jasper Goedbloed (1992) on how to avoid crosstalk is:

- Give each circuit loop its own conductors.
- Keep the two conductors of a loop as close as possible.
- Use no frequencies higher than strictly necessary.
- With printed circuits, use two-sided or multilayer boards.

Kitchen gas lighter as a test probe

One source of man-made noise is of interest because it can be used as a testing tool. A piezoelectric gas lighter held close to a pocket calculator will change the numbers shown on the display when it is switched on. A correction can be made only by switching off and starting again. If, as soon as the first printed-circuit board is built, the most susceptible digital circuits of a PCB are tested in this way, it is possible to correct them without too much expense and delay.

Coupling channels

Any electronic interference reaches the victim device through a 'coupling channel'. This always invisible route may involve electrical conduction along a wire or some sort of radio reception or magnetic induction. Interference can be reduced or eliminated if the coupling channel can be blocked.

Electronic interference can be reduced by several methods alone or in combination, including distance, shielding, grounding (earthing), tuning (matching or balancing), filtering, orientation and finally electronic redesign of the victim or the source (emitter). Distance and orientation have been briefly mentioned in connection with the portable radio placed next to an active computer. The first four are dealt with in separate chapters of this book.

Warships, often with over 100 antennas transmitting or receiving, probably suffer more interference than any other modern environment. To reduce interference the US Navy therefore introduced plastics or timber for railings and other structural units. But it has discovered that even these can obstruct reception or transmission to such an extent that they must be placed outside the angle of vision of an antenna.

Decibels

Comparisons, for EMC purposes of noise or signal strengths, are made with a unit called the decibel (dB). It expresses any ratio of two values, whether powers, field intensities, currents, voltages, etc. Although only a tenth of the bel, it is still a large unit and so is convenient for dealing with large ratios; in fact a ratio of 1 000 000 reduces to 120 dB for current or voltage or 60 dB for power. But being logarithms, dB are not easily converted without Table 1.3 given below. The dB values for voltage or current are twice those for power so one needs to be sure which is being discussed.

The meanings of dBm and dBW need explaining. The dBm values are calculated by comparison with one milliwatt (mW), so that 2 μW (microwatts) are equivalent to −27 dBm (minus 27 dBm): dBW implies dB in relation to 1 W (watt), see Table 1.4.

Fractions of a decibel should not be used, being practically meaningless even if they are printed in hi-fi catalogues.

The future

However happy we may be that the EMC law has come into force, there are people who are deeply dissatisfied with its lack of effectiveness in curbing the harm done by some types of EM waves to animals. Morgan (1994) points out: 'The biological effects of exposure to

Table 1.3 Conversion of ratios of power, current or voltage to decibel (dB) values

Ratio	dB values of voltage or current	dB values of power
1 000 000	120	60
100 000	100	50
10 000	80	40
1000	60	30
100	40	20
10	20	10
9	19	10
8	18	9.0
7	17	9
6	16	8
5	14	7
4	12	6
3	10	5
2	6	3
1	0	0
0.1	−20	−10
0.01	−40	−20
0.001	−60	−30

Adapted from C. R. Paul, *Introduction to EMC*. Wiley, 1992.

Table 1.4 Conversion table between dBW and watts

Watts	dBW
1000	+30
100	+20
10	+10
1	0
0.1	−10
0.01	−20
0.001	−30

electromagnetic fields are likely to become a significant technical, legal and social issue in the 1990s.'

Among the objectors are some who claim that their health has suffered severely because of the electromagnetic waves emitted by their own appliances or by overhead or underground electric power cables. Some of their difficulties are described in Chapters 13 and 14. Powerful electric power companies in all countries issue denials.

Chapter 2
Regulations and enforcement

The always chaotic but alternately tragic and ludicrous mishaps related in Chapter 1 forced European experts in 1989 to insist on the stern EU directive 89/336/EEC with its threat of criminal prosecution. Shock at the chaos, felt equally early in the USA and Germany, is seen in the regulations both of the FCC (Federal Communications Commission, USA) and of the VDE (German electrical association) which are equally strict, if not more so. The EU sensibly insists on the need for unification of EMC standards. Most countries agree, so national standards generally match the guidelines laid down by the EU. The importance of EMC becomes evident if we recall that modern aircraft no longer use manual controls. Control is by computer. Cars have gone the same way.

Designers interested in EMC often have difficulty in finding out which regulations apply to the product in question, especially if the product is to be sold in different countries. In Britain help may be obtained from the IEE (Institution of Electrical Engineers) or the BSI (British Standards Institution) or NAMAS or a NAMAS-approved laboratory or from an EMC consultant, although often subject to a fee. But a manufacturer's trade association can usually submit queries fairly forcibly to the BSI since most manufacturers pay heavy BSI subscriptions. In 1995, after their move to Chiswick, in west London, the BSI library were asking non-members of BSI to pay a standard £25 for half-a-day's reading.

International bodies

A short explanation of the main international organizations concerned with EMC may help. The IEC (International Electrotechnical

Commission) aims to set world standards and works closely with ISO (International Standards Organization). Both are in Geneva. The International Special Committee on Radio Interference (CISPR) is the IEC committee responsible for EMC. Two organizations, concerned only with Europe,cooperate closely with IEC, their 'world parent'. These are ETSI (European Telecommunications Standards Institute) and CENELEC (European Organization for Electrotechnical Standardization). Some abbreviations for the names of interested bodies are listed at the end of this chapter.

The IEC's electrotechnical vocabulary (IEC 50) has definitions in three languages: English, French and Russian, accompanied by corresponding terms in Dutch, German, Italian, Polish, Spanish and Swedish. Some of the English sections correspond to some of the 50 or so parts of BS 4727.

An important IEC publication for EMC is IEC 801 which is published in several parts that cover different aspects for the compliance of electronic units with the EMC directive. IEC 801-2 covers testing for ESD (electrostatic discharge). IEC 801-3 defines the levels of immunity to radiated EMI required in equipment for industrial process measurement and control.

Another important IEC publication for EMC is IEC 1000, which is made up of many parts that are published separately as they become ready. Some 85% of IEC standards are published verbatim as British Standards (BS). An EN prefix to a BS means that it is a European Standard, 'Norm' and 'norme' being the German and French words for 'standard'.

Some prefixes to titles of regulations

CFR	Code of Federal Regulations, from the FCC in the USA.
DD	Draft for development (BSI).
FCC	US Federal Communications Commission.
HD	CENELEC harmonization document.
OJ	Official Journal of the EU.
prEN	Provisional (draft) standard.
prETS	Provisional ETSI standard.
SI	Statutory Instrument, a UK law. The 'SI' precedes a year and another number.
Vfg	(Verfügung) a decree of the VDE, the German electrical engineers' association.

The British Standards Catalogue, published every year, contains much information on international or European standards. Because of the need for international agreement, it often happens that British standards (BS) acquire new prefixes as well as new numbers. For instance BS 800, on radio interference, has become EN 55014 but is still a BS. Its new number shows the reader that the BS now has validity throughout Europe. Its last two digits, '14', show that it comes from CISPR 14.

A European standard, drawn up to help in achieving EMC, is EN 55022, also known as BS 6527. Its final numbers, '22', indicate its origin in CISPR 22. It specifies that between the frequencies of 30 and 230 MHz the electric field strength emitted by digital equipment, measured 10 m away from it, must not exceed 30 μV/m (microvolts per metre). Rule-of-thumb tests show that equipment usually exceeds this limit when common-mode current (stray current, explained in Chapter 7 and in the Glossary) in a connecting cable is more than 5 μA (microamps).

Standards which demonstrate compliance with the EMC law must have been published in the Official Journal (OJ) of the European Communities. Harmonized standards are those which have been so published, although they may, in the UK, have a BS prefix. The English BS text will be the approved version of the EN in the other European languages. Compliance with the EMC directive is required in all countries of the EU as well as of the European Economic Area.

Generic standards

To fill gaps where no standard exists, CENELEC are developing a range of 'generic' standards of wide application. These do not introduce new tests, nor are they concerned with any particular type of product. They describe an environment, e.g. domestic or heavy industrial, aiming to cover the requirements of the EU's EMC law. EN 50081 Emissions and EN 50082 Immunity were the first of these generic standards. If other standards exist for the appliance concerned, they take precedence over the generic standards. Some 'basic' standards of the IEC which are referred to in the generic standards are the different parts of IEC 801 and IEC 1000. Like most other standards, all those referred to in this paragraph have several parts.

Chris Marshman's *Guide to the EMC Directive 89/336/EEC*, 2nd edn 1995, helps to disentangle the complexities of the law and its regulations. Marshman is an electrical engineer who in 1984 established the EMC consultancy and test facilities in York University and runs its EMC courses at MSc and lower levels. He also produced the Institution of Electrical Engineers' video package that teaches EMC. Unfortunately his book, like most of those written by electrical engineers, is not easy to read but the obstinate effort and time needed to read him is only a small fraction of what would be needed to read and absorb the permanently changing regulations and standards, written in 'legalese', which is even stranger.

Test laboratory

If the client does not specify the test house in his contract, the manufacturer may have difficulty in finding a suitable one. Not every laboratory can test large units. 'Small' units are no larger than one metre cube. Test laboratories have to be booked several months in advance and the manufacturer must include the prospective test laboratory in his test plan. A manufacturer missing a date booked with a test laboratory can suffer severe delay in getting a product on to the market. Another test laboratory may even have to be found. Severe shortages of test facilities are forecast for the next few years, not only in the UK.

The consulting engineers W. S. Atkins produced a report, published by the Department of Trade and Industry in 1989, describing EMC test facilities available in the UK and comparing them with the demand. Their count of the test facilities attained only one-sixth of the demand. The situation for open field test sites was even worse. For these the demand was twelve times as high as the number available. Atkins estimated that the backlog of testing would not be cleared for 15 years and that the worst bottleneck was the shortage of EMC engineers. Moore's more recent (1993) survey, though much smaller, may be closer to the current situation.

Conformance testing

Conformance testing is the final testing procedure, showing the client that the appliance satisfies his requirements. The much earlier preconformance testing should have shown how to eliminate any problems. But the conformance testing can also be the basis for marketing as well as providing the manufacturer's necessary background in the event of a dispute about the product or its EMC.

A well-equipped laboratory with large, semi-anechoic rooms for EMC testing can cost £1 million to build but sometimes may be hired for a few thousand pounds per day. A screened semi-anechoic room is difficult to build for frequencies below 200 MHz. Such rooms have cones or pyramids of RAM (radio-absorbent material) fixed over the walls and ceiling to absorb radio waves and prevent them from being reflected. Semi-anechoic or 'screened rooms' usually have RAM on the walls and ceiling but not on the floor.

Conformance testing for radiated emissions should be done on an open-area test site (OATS) shown in Fig. 2.1. To avoid echo effects, no

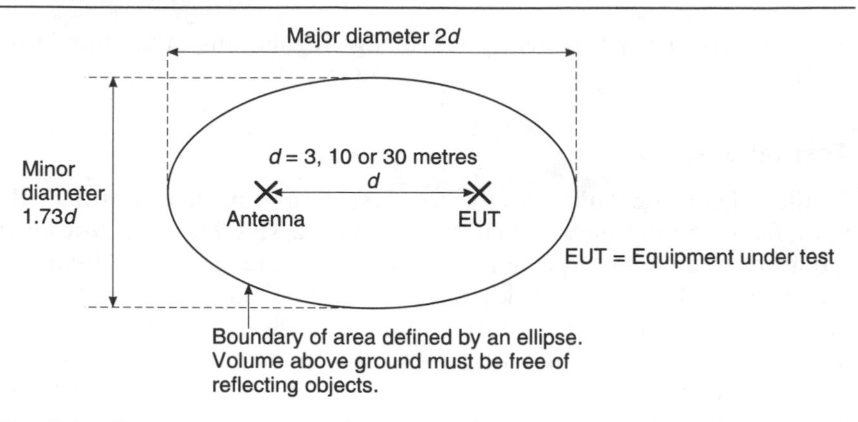

Fig. 2.1 Open-area test site (OATS) conforming to EN 55022 (CISPR 22)

structures, overhead wires nor other objects that might reflect RF waves are allowed in the clear area of the site, and preferably also not near its boundaries. A minimum OATS is defined in recent editions of CISPR 16 and in EN 55022. According to circumstances the distance *d* between the equipment under test (EUT) and the measuring antenna may be 3, 10 or 30 m as the client chooses. The clear area is an ellipse with diameters 2*d* and 1.73*d*. If tests are made with *d* at 3 m, it is assumed that levels with *d* lengthened to 10 m will be 10 dB lower.

A 30 m open-field test site needs at least an area of 60×52 m to avoid any possibility of reflecting metal such as parked cars outside the site. Such an area in or near a city can be very expensive. To counteract the effects of bad weather, some sites are provided with a radio-transparent building often made of GRP (glass-reinforced polyester). To avoid disturbance to the measurements the control room may be in a basement below the site. One GRP radio-transparent structure measures 20×12 m and covers the entire ground plane.

Radio reflections from the ground cannot be prevented, but they can be made repeatable by placing a ground plane (a metal or wire mesh sheet) under both the antenna and the EUT, projecting 1 m beyond the EUT all round. Ideally joints in the metal ground plane (Fig. 2.2) should be welded continuously.

A metal ground plane is helpful because it allows measurements to be more repeatable, theoretically enabling the amount of RF reflection from the ground to be the same at every test. Soils in different places have different conductivities, therefore different abilities to reflect RF waves. The most highly conductive soils, with the best reflection, are usually the wettest. If mesh or perforated metal is used as a ground plane the holes should not be larger than one-tenth of the wavelength at 1 GHz which is 30 mm.

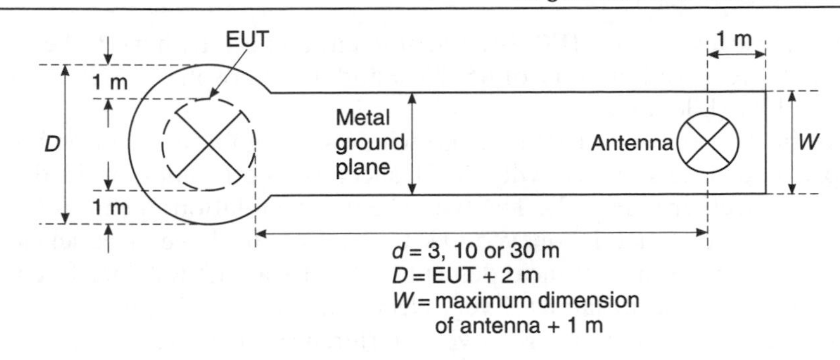

Fig. 2.2 Minimum size of metal ground plane in an OATS for radiated emission testing to EN 55022. EUT = equipment under test (maximum dimension)

In Fig. 2.2, the diameter, *D*, of the ground plane around the EUT is equal to the greatest dimension of the EUT + 2 m. The width *W* of the ground plane along most of its length is equal to the greatest dimension of the antenna + 1 m.

Screened rooms and OATS

A screened room or an anechoic room has two advantages over an OATS. It excludes the external environment, and enables interference with other laboratory equipment to be avoided. It is also much smaller than an OATS, but it has several disadvantages:

- The size of the EUT is restricted because it depends on the room size.
- The room has reflections and resonances because it is metallic.
- Anechoic rooms can be prohibitively expensive.

A screened room, sometimes called a Faraday cage, may be easier to find than an OATS but to make it anechoic is expensive and when properly done, the RAM pyramids on all the walls and the ceiling, at least 1 m long, use up much of the room's volume. (At 30 MHz they need to be 2.5 m long.) A screened room shields sensitive equipment against incoming radiation, as well as preventing interference travelling out to disturb other equipment. Its area will be at least 4 × 5 m (× 3 m high) to allow 1 m beyond the 3 m minimum test range. However if the current revision of IEC 801-3 is approved, which seems likely, these dimensions will have to be enlarged to 6 × 5 × 3 m, implying a considerable increase in capital investment.

For most tests under IEC 801 the ambient conditions have to be 15–25°C, at a relative humidity of 45–75% and an atmospheric pressure of 68–106 kPa (kilopascals).

Normally a screened room has no windows. A good one can be built of plywood sheets plated with steel sheet on both sides, welded or clamped together (Fig. 2.3). The two essential ventilation openings (one in, one out) use metal honeycombs to cut off the lower frequencies. Lighting is by ordinary incandescent bulbs to avoid the interference created by fluorescent lamps. Every electrical cable coming in has to be connected to a filter to remove interference. The access door is important and should have a beryllium-copper finger strip all round the frame.

For the very highest protection, such as that needed against a NEMP, the steel sheet would be plated all over with copper sheet. The slots at all joints would also be covered by copper tape. But, as stated earlier, the EMC law does not impose a requirement for NEMP protection.

Lining with pyramidal absorbers is expensive, doubling the cost compared with an unlined room, as well as reducing the room size appreciably. The pyramids are made of carbon-loaded foam but they do not function well below 200 MHz. An alternative to lining the walls with RAM absorbers is to line them with ferrite tiles or ferrite grids but

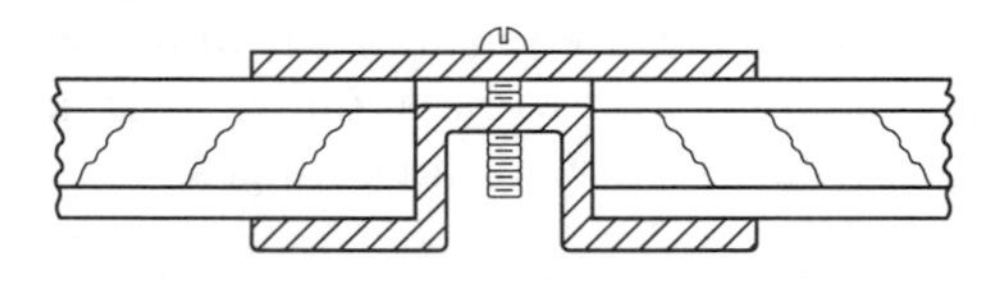

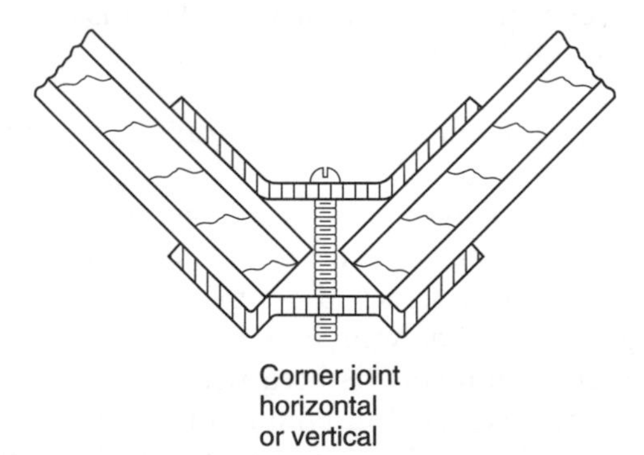

Fig. 2.3 Screened room: typical joints between wall panels of plywood faced with steel sheet both sides (after Violette *et al.*, 1987)

these are also expensive and not easy to fix, although they occupy less space.

The screened room usually needs access to two neighbouring screened rooms which contain computers and other devices. These emit radiation that is excluded from the main test room by its screening. But because of their metal construction, other errors in measurement known to occur in screened rooms can result from the reflections or standing waves in them.

Test plan

Every unit being tested should have its own strict test plan. It may be drawn up by the testing laboratory as part of its contract or by an EMC consultant, or by other EMC specialists. The plan must be written out well in advance of testing and, where feasible, submitted for approval to the customer. In some contracts the test plan is subject to guidelines laid down in the agreement. For example, the test plan for a unit made to the American military standard MIL STD 461 has to be written according to D1-EMCS-80201, while tests have to conform to MIL STD 831.

In the USA, the FCC has published rules to reduce interference, entitled CFR 47 (Code of Federal Regulations). Part 15 of CFR 47 refers to unintentional broadcasts of interference and basically concerns digital devices such as computers using timing signals at 9 kHz or faster. The two main classes are:

- class A for business or industrial use,
- class B for domestic use.

Class B is the stricter because a TV or radio could be severely affected by interference from a nearby guilty computer. Publication FCC/OET MP-4 describes, for computers and similar units, how to measure the interference they emit.

European authorities use ‘A’ and ‘B’ in the same way. A domestic appliance (to Class B) is usually understood to be one which can be satisfactorily worked from a 13-amp socket outlet.

Repeatability of tests

For several reasons, the results of EMC tests are not easily repeatable. First it is far too time-consuming and expensive to use a wide range of frequencies to determine the most appropriate measurements of the maximum radiated field strengths of any particular unit. Test results consequently are far less accurate and reliable than those made for common quantities such as weight or direct current or voltage.

Measurements for EMC of conducted electric current can be made with a repeatable accuracy of 6–10 dB but measurements of radiation can depart much further from each other, often by more than 20 dB. This is why tests must be made in a disciplined way, following an agreed plan, and test reports must give all details of the layout. The positioning of a cable as well as its length and type can make a big difference to the test result, so these details should be carefully recorded in the hope of achieving repeatability.

Field strengths

As might be expected, electromagnetic field strengths decrease with increasing distance from the electromagnetic source. In built-up areas the decrease is much more rapid than in open country. Some estimates make the decrease in built-up areas proportional to the cube of the distance, whereas in open country it is roughly proportional only to the distance. Field strengths for good radio and TV reception may be only a few microvolts per metre (μV/m). But at more than a million times this level, an airport radar search-transmitter produces an electric field estimated at 3–120 V/m at a distance of a few kilometres from the airport. A long-wave broadcasting transmitter has a field strength of 5–10 V/m, also a few kilometres from the source. Chapter 14 has more on field strengths.

A photographer's electronic flash bulb, 1 m away, emits 60 V/m for an amateur model and 150 V/m for a professional model with a 300-joule flash lasting about 50 μs. The London underground railways therefore forbid them, to avoid disturbance to train drivers and to the electronic signalling for trains.

In the future as the cost of manpower rises and computer modelling becomes cheaper, it is possible that computer simulation may be easier to achieve than test measurements. At the moment, however, test measurements have to be made as reliable and repeatable as possible.

Declaration of compliance

Three possibilities

The EMC law, Directive 89/336/EEC, requires every manufacturer or importer of electrical or electronic units into Europe to make a 'declaration of conformity (compliance) with the law' for the unit in question. A 'CE' must also be fixed on the unit in letters not less than 5 mm high, legibly and indelibly. The declaration can be based on published standards alone, if they exist. Such a declaration of 'self-certification' is easy, simple and involves no delay. But every unit must carry the letters ‘CE’. If relevant standards do not yet exist, and this often applies to ultra-modern products, there is a second possibility. The importer or maker must base the declaration on a ‘technical construction file’ for the product. In Britain, the file must be approved by a NAMAS-accredited laboratory (a ‘competent body’), involving payment of a fee and a delay during the laboratory investigation.

NAMAS accreditation, however, does not prove that the accredited laboratory has the technical ability to do the job. According to Marshman (1995), the accreditation is only an indication of quality assurance to BS 5750. Some manufacturers hope to obtain NAMAS approval as competent bodies, and this, if achieved, will certainly speed up the entry of their products into the market.

Very few products are wholly described by existing standards. Allowable radiation is prescribed more often than allowable immunity. Usually the FCC, unlike the EU, states few requirements for immunity. The only products in Europe covered by both immunity and emission (radiation) standards are TV or radio receivers. If appropriate standards do not exist, a manufacturer can use a mix of product-specific and generic standards to self-certify the product.

In the UK, having built the equipment to NAMAS requirements, the maker prints ‘CE’ on it to demonstrate this official approval. In Germany, there are wide-ranging, highly-qualified and officially-approved consulting engineers called TÜV (Technischer Überwachungs Verein) who approve or disallow everything from the roadworthiness of a car to the structural safety of a tall building. The TÜV mark printed on equipment (Fig. 2.4) relieves the maker of any responsibility in a law court because the onus of proving its dangers or faults now lies on the user and involves high-level engineering arguments with the local TUV, which only a brave, highly-qualified engineer would dare to undertake.

For radio transmitters and transceivers, other than amateur equipment, there is a third route to compliance with the law. This involves approval for EMC by a notified body, which in Britain is the BABT (British Approvals Board for Telecommunications). Whichever route is taken, the declaration of compliance (conformity) must be made and the documents kept available for ten years from the first entry to the market.

For telecommunications terminal equipment such as home telephones, fax devices, radio transmitters, transceivers, etc., there are special regulations in addition to the EMC law. These special regulations have always existed in the UK and other countries because devices connected to the telephone can affect the working of other telephones. The BABT is therefore still the approving authority for them even though the regulations are now becoming European rather than British.

Fig. 2.4 TÜV trademark in Germany (after Morgan, 1994)

Failure to comply with the law makes the appliance unsaleable in the EU and renders its manufacturer or importer liable to criminal prosecution. In Britain, prosecutions are the responsibility of the Department of Trade and Industry. Its office that is concerned with radio, TV, telecommunications and EMC is the Radiocommunications Agency (RA), Waterloo Bridge House, Waterloo Road, London SE1 8UA. The situation in 1996 was that prosecutions by the RA would be initiated only on the basis of complaints by the public.

From the viewpoint of the EMC law each country has three layers of authority:

1. The competent authority, which in the UK is the Department of Trade and Industry.
2. Competent bodies, authorized by NAMAS to operate a test laboratory.
3. Notified bodies, identical to (2) but in addition authorized to issue a certificate of approval of an appliance used for telecommunications.

Some of the delay and costs involved in building up the technical construction file can be avoided if the manufacturing company tests the main components in its own laboratory, even if this is not, as it should be, a screened anechoic room as described previously. In rigorously comparable conditions, important components for which standards exist can be tested, choosing the component with the best performance.

There are four levels at which the law can be broken:

- in the system as a whole,
- in a subsystem or unit of equipment,
- in a printed-circuit board,
- in a component, e.g. transistor, integrated circuit, etc.

At frequencies below 30 MHz the authorities require the magnetic field strength to be measured. At higher frequencies the electric field is of interest.

Two loop-antenna probes (search coils) which can test the legality of a device are shown in Figs 2.5 and 2.6. Figure 2.5 is heavy and expensive with a loop of 60 cm diameter but is approved by the authorities. Figure 2.6 shows a very much smaller, inexpensive home-made loop antenna which can be used for the lowest frequencies that cannot be measured by the large probe.

History

The International Telegraph Union or ITU (since 1992, the International Telecommunications Union) was founded in 1865 during the first rapid growth of the telegraph system. The inaugural International Radiotelegraphy Conference in Berlin in 1909 decided on the 500 kHz frequency and SOS signals to be used by ships in distress and those rescuing them. The ITU joined the United Nations after 1945 and its member states developed regulations governing the use of frequencies

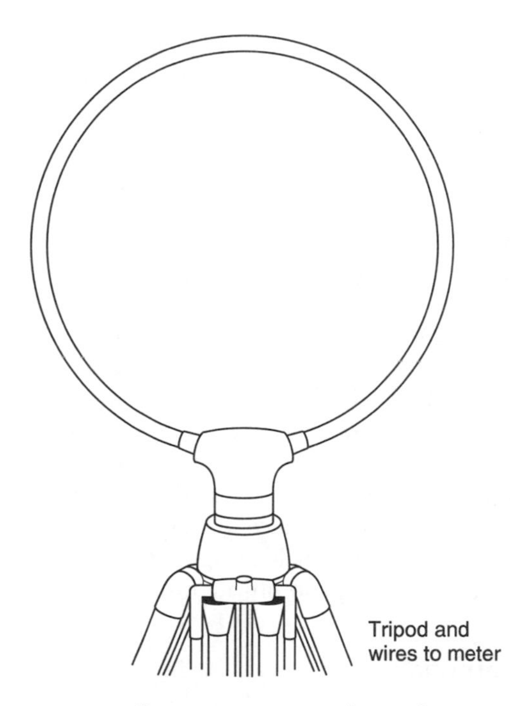

Fig. 2.5 Loop antenna for statutory testing of magnetic field (after Goedbloed, 1992)

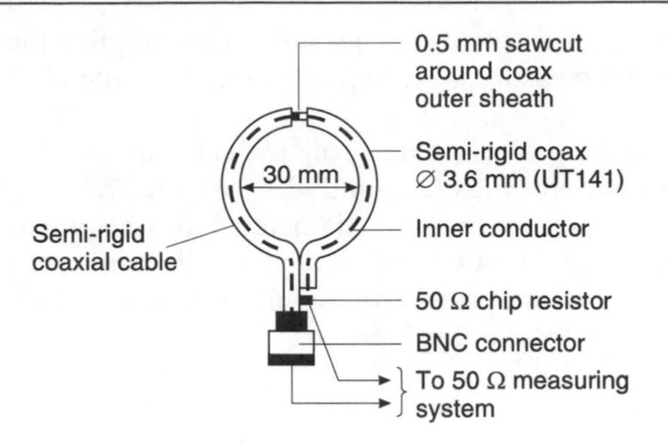

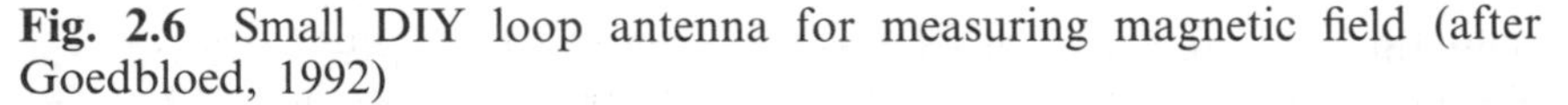

Fig. 2.6 Small DIY loop antenna for measuring magnetic field (after Goedbloed, 1992)

and the reduction of interference. The IFRB (International Frequency Registration Board) is the part of the ITU that decides who shall use which frequency for broadcasting.

But evidently what the ITU did was not enough to satisfy the CISPR because it came into existence in 1933 in Paris as a mixed committee of the IEC and the IBT which was deeply concerned about radio interference from electrical equipment. The CISPR is now a committee of the IEC. Other international organizations interested in EMC are listed on the last page of this chapter.

Enforcement

Member states of the EU have to make sure that equipment does not reach the market if, although carrying the 'CE' mark, it does not comply with the law, 89/336/EEC. If it has already reached the market, the country must ensure its withdrawal. Without evidence of such noncompliance, member states have no right to impede its free movement in their market. But buyers or retailers of an appliance cannot be expected to know whether or not something is illegal.

Because of these difficulties, the UK authorities will not immediately enact a criminal offence of using or selling noncompliant appliances. Related criminal offences which already exist in the UK include misuse of the CE mark, false or misleading information to a competent authority, and breach of a prohibition notice.

Testing for compliance

There are four internationally-accepted classes of test, listed below. For each class there are devices that verify conformity.

1. conducted emission,
2. conducted susceptibility,
3. radiated emission (radiation),
4. radiated susceptibility.

At frequencies below 30 MHz, according to the regulations the magnetic field strength has to be measured, often with the loop antenna of Fig. 2.5, at the prescribed distance of 3, 10 or 30 m from the EUT. In an open-area test site the EUT is normally protected from the weather in a wooden hut. The wood does not reflect EM waves if it is dry. To make sure that no metal (therefore radio-reflecting) objects are nearby, the measuring equipment, computer and its peripherals are underground or well away from the open-area test site. The EUT is rotated until its radiation maximum has been found. The antenna is then rotated to find a maximum. The main difficulty, according to J. Goedbloed (1992), is that ambient disturbances, mainly from distant and perfectly legal transmitters, are generally well above the maximum allowable limit for the EUT.

Management plans for large units

Serious problems occur in achieving EMC compliance for units as large as textile machinery, railway rolling stock, telephone exchanges, or electrical generating equipment, because they are large and unique. There are also usually no standards for them. Because they are not 'placed on the market' but 'taken into service', some do not need to carry the CE mark but nevertheless must be shown to be EMC compliant, generally by submission of the usual technical construction file to a 'competent body' – a test laboratory approved by NAMAS.

At the start of the project an EMC Management Plan should be drawn up identifying possible sources of interference and listing texts such as standards, consumer specifications, etc., explaining the theory of the plan. The EMC control of suppliers may involve each of the suppliers submitting their own EMC plan. The purpose of the EMC Management Plan should be explained, giving dates for achieving particular EMC targets. Suppliers' subsystems should be identified and each supplier should be asked to provide documents describing the EMC compliance of their subsystem.

The maker should choose the competent body early in the manufacturing process because it will define the tests to be reported in the construction file. The Department of Trade and Industry possesses a list of competent bodies with their specialities, which can be made available to manufacturers.

Some help is given by an ETSI provisional standard, prETS 300 127,

'Radiated emission testing of large telecom systems', which states that on a 30 × 30 m open-field test site the radiated emissions of a unit measuring up to 6 × 3 m may be measured by an antenna that moves around it at a minimum distance of 10 m. So far as immunity testing, whether conducted or radiated, is concerned, this may be possible only at the completed installation or by splitting the unit into different components, enabling each to comply separately.

Usually many different frequencies are involved in the testing of large systems. Immunity to power frequencies (50 or 60 Hz) is dealt with by IEC 1000 Part 4 Section 8. Other EMC testing is explained in IEC 1000 Part 2 Sections 4 and 5.

But large systems also use high frequencies and these involve transients and radiated EM fields. IEC 801 Parts 3, 4, 5 and 6 are relevant. IEC 1000 Part 4 Section 4 is mentioned at length in Marshman's book (1995).

Measurement units: power density and specific absorption rate

Power density

Below the frequency of 30 MHz, electric fields (E) are measured in volts, millivolts or microvolts per metre (V/m, mV/m, or μV/m) and magnetic fields (H) in amps, milliamps or microamps per metre (A/m, mA/m or μA/m). At higher frequencies above 30 MHz, the power density (power flux density) is more usual, measured in watts per square metre (W/m^2) or, in the USA in milli- or microwatts per square centimetre (1 mW/cm^2 = 10 W/m^2). Power density is the product of the two fields, E and H, E being in V/m and H being in A/m, (see Chapter 14, p. 177).

As an assessment of biological exposure, magnetic field strength is usually measured by the milli-, micro- or nanotesla (thousandth, millionth or thousand-millionth of a tesla). One tesla = 10 000 gauss, the older unit. (More on field strength units in Chapter 14.)

Specific absorption rate

For the purpose of simplifying comparisons of results on human or animal bodies in a way that is more accurate than power density, the specific absorption rate (SAR) was invented. It is the power applied in watt, per kilogram of tissue or of total body weight. There are two types of SAR, one based on the whole-body weight, the other based on the weight of the local tissue under discussion.

Other more complex methods of measurement are mentioned in the

WHO report (1993), *Environmental Health Criteria (EHC) No 137. Electromagnetic Fields, 300 Hz to 300 GHz*. If used cautiously and with very strong reservations, the SAR may sometimes allow results on animals to be transferred to humans.

According to R. Kitchen (1993), the SAR is relevant for frequencies between 0.1 MHz and 6 GHz. Above 6 GHz, power density limits are used to control exposure to radiation. At lower frequencies below 100 kHz the limiting factors are the current density in the human body and any electrical stimulation of tissues.

Instruments for measuring field strength

In his *RF Radiation Safety Handbook* (1993), Kitchen describes instruments for measuring electric and magnetic fields from the viewpoint of the safety of nonionizing radiation (NIR). Such instruments are concerned mainly with higher frequencies. Though 'RF' excludes X-rays they do exist in transmitters or other devices, such as cathode ray tubes and older sets that contain valves, because these use some tens of thousands of volts. Wherever electrons travel under such high voltages, X-rays may be produced, although they are usually well shielded and kept safely inside an appliance, as in a TV screen.

Every field-measuring instrument includes a meter and a sensor. The sensor contains an antenna and a detector. The detector is either a thermocouple or a diode circuit, either of which can be easily overloaded and burned out. The antenna is normally a probe on the end of a tube forming the handle. At, for example, 1 GHz the wavelength of 30 cm does not create difficulty for the handling of such an antenna and at higher frequencies the wavelength is even shorter. The best instruments are isotropic – those which can measure fields of any polarity in any direction. Nonisotropic instruments such as the loop antenna of Figs 2.5 and 2.6 are cheaper but do not measure the total field, only that from one direction. Some instruments measure the wide range 0.3–40 GHz but only the electric field. Below 0.3 GHz (300 MHz) the magnetic field also may be measured.

Power density is the commonest unit used in RF safety measurements but it exists only for plane waves, i.e. those in the far field. Strictly speaking, no instrument exists yet which can measure it because they normally measure only one of the two variables, E and H – usually the electric field, E.

Instrument overload

Any type of sensor can easily be overloaded, resulting in the disaster of burnout. Even walking around with the instrument switched off can cause an overload. Consequently any instrument not in use should be shielded, either by keeping it in its metal case or wrapping it in aluminium cooking foil, which may be almost as effective.

An alarm circuit is therefore a valuable feature of any field-measuring instrument. It sounds a warning when the full-scale deflection becomes imminent. It should also sound continuously if the probe suffers burnout or open-circuit because of a break in the cable etc. In a noisy environment the alarm may sound but be inaudible. For this reason another valuable feature is 'maximum hold', an indication of the maximum value experienced.

Before buying any instrument ask to see the accompanying instruction book, read it carefully, and check that the instrument has essentials such as:

- Inbuilt battery charger,
- Battery voltage check device, or low-voltage display,
- Is the chosen instrument properly shielded by a metal carrying case?

Rechargeable batteries do not last for ever but eventually need replacing.

Handling of instruments for measuring EM fields

A drifting zero is a common source of wrong readings. All instruments have drifting zeros and must be adjusted to bring them back to read correctly. Normally the instrument has to be removed from the field of measurement although some makers claim that their instrument can be adjusted in the presence of a field. It may be possible to exclude a field from a sensor by wrapping it completely in aluminium cooking foil.

Calibration includes setting the zero correctly and this can be done by any reputable instrument maker.

Possible wrong readings

A magnetic field (H field) sensor should respond only to an H field. Similarly an electric field (E field) sensor should respond only to an E field. But if the H field is small, a strong E field may affect the H field meter reading.

At lower frequencies, below 1 MHz, the lead to the meter may act as an extension of the antenna, picking up RF and giving wrong readings. (At 1 MHz the wavelength is 300 m.) Wrong readings may also come from a strong signal outside the bandwidth of the instrument. One such strong frequency can affect the meter reading. If possible, switch off what you believe is the possible source. If the disturbance disappears you will have found the source.

A useful check on whether a reading is false can be made by covering only the sensor area of the probe with aluminium cooking foil and then trying for a second reading. The cooking foil should be kept out of contact with the earth, the meter and the rest of the probe. If the meter reading is unaffected by the cooking foil, then it is likely to be false.

Waveguide dangers

Kitchen explains the great dangers that exist both for the instrument and for the operator, when a probe is inserted into a waveguide. Replacement of an instrument is expensive, in addition to the costs incurred by the absence of the instrument while it is being repaired. The presence of the operator holding the probe can significantly disturb the meter reading. In the near field in any case, all measurements must be made in the operator's absence. Some writers therefore affirm that to obtain a correct reading, the probe containing the antenna should be at least 2.5 m (8 ft) away from the meter. This can be achieved by holding it extended forwards on a wooden pole 9 ft long. The meter can be held at the operator's side where it is easy to read. An isotropic probe, normal for most work, should be aimed towards the source and rotated through 360° to obtain the peak isotropic response. This method is difficult with a probe that is rigidly plugged to the meter because rotation through 360° is not possible without completely obscuring the meter, although this rigid fixing is, however, convenient for an operator who must work single-handed.

In an instrument with the 'maximum hold' feature, this should be switched off as soon as it has been read so as to avoid confusion later.

Electrostatic charges can build up on instruments, often because of the presence of a plastic cover on a document, file or book. Such charges can be a nuisance because they create transient readings which only disappear slowly and patience is needed to allow for the slow discharge.

In practice, the repeatability of measurements of EM fields is extremely poor, shamefully so by the standards of most scientists and engineers. Even with good calibration the uncertainty is from ±30 to ±40%. Consequently, when approaching a statutory limit, an allowance of at least ±30% should help to avoid exceeding the limit.

Crime with stolen mobile phones

Cloning fraud at £3000 per day

Criminals who have acquired numerous analogue mobile phones can gross £3000 a day by hiring them out in pubs or clubs for £20/hour each. But by the end of 1996 they had not been able to unlock the secrets of the GSM mobile phones which are used all over the world including the continent of Europe, although it was expected that they would do so soon. The older analogue phones already bring them rich pickings.

The FCS (Federation of Communication Services), representing the four British cellphone operators, Cellnet, Mercury, Orange and Vodaphone, complains that the criminals are winning and that the cellular phone industry is losing £200 million a year in fraudulent calls. Since 1994 the FCS and the operators have been pressing Parliament to make it illegal to advertise, sell, own or use cloning equipment. In the USA and Hong Kong such laws exist already.

'Cloning' means to give a stolen phone the identity number of a legitimate mobile phone subscriber. In this way expensive calls to the continent of Europe, South Africa or New Zealand can be made at the expense of victims whose numbers have been stolen. A scanner is a special radio receiver which can search for particular wavelengths and does so much more quickly than humans can. With a specially-programmed scanner, criminals can record on a computer the secret numbers of legitimate mobile phones and transfer these numbers to their stolen phones. Criminals prefer to listen for the secret numbers near to an airport because this may give them the advantage of the absence abroad of the legitimate owner of the stolen number for several days.

The re-programming needed to transfer these numbers into stolen mobile phones is not simple but instructions on how to do it are available on the Internet. Although a legitimate owner normally notifies his exchange that his phone has been stolen, he may not immediately realize what has happened. In these few hours the criminal can charge him up with costly trans-continental phone calls to the owner's account.

Bodies concerned with EMC documents

ACEC	The Advisory Committee on EMC of the IEC
AFNOR	French standards organization, Paris.
ANSI	American National Standards Institute
ASTM	American Society for Testing and Materials
BABT	British Approvals Board for Telecommunications
BECA	British Electrostatic Control Association
BNCE	British National Committee on Electroheat
BSI	British Standards Institution
CCIR	International Radio Consultative Committee
CENELEC	European Organization for Electrotechnical Standardization
CIEH	Chartered Institute of Environmental Health, formerly Institute of Environmental Health Officers
CISPR	International special committee on radio interference, now the EMC committee of IEC
DIN	German National Standards Organization, Berlin
ECMA	European Computer Manufacturers' Association
EIA	Electronic Industries Association (USA)
EOS/ESD	Electrical Overstress/Electrostatic Discharge Association. 200 Liberty Plaza, Rome, NY 13440, USA
EOTC	European Organization for Testing and Certification
ETSI	European Telecommunications Standards Institute
FCC	Federal Communications Commission (USA)
IEC	International Electrotechnical Commission
IEE	Institution of Electrical Engineers (UK)
IEEE	Institute of Electrical and Electronic Engineers (USA)
IFRB	International Frequency Registration Board
ISO	International Standards Organization
ITU	International Telecommunications Union
MIRA	Motor Industry Research Association
NAMAS	National Measurement Accreditation Service (at the NPL)
NPL	National Physical Laboratory, Teddington
RA	Radio Communications Agency of the UK Department of Trade and Industry
SAMA	Scientific Apparatus Makers' Association
TÜV	Technischer Überwachungs Verein, official engineering consultancies in Germany, with wide responsibilities
URSI	International Union of Radio Science
VDE	German electrical association (Verein deutscher Elektriker)

Chapter 3
Power supplies

Batteries are an unusual power supply because they bring almost no interference. Unfortunately all batteries have electrical resistance and this has associated noise, although it is not very intense. But they are heavy, expensive, and need periodic recharging so most people prefer mains power if it can be suitably organized. Computers need low-voltage DC, so batteries are sometimes essential for them, although even here, where possible, mains AC power is preferred. It can be easily rectified to DC. Chapter 10, Transients and protection against them, reviews the subject of AC power supply voltage from the more specialized angle of making transients safe.

The mains power supplies are notorious generators of noise. The mains voltage fluctuates with switching on or off and with varying load. Usually it is imperative to reduce the random noise from the semiconductors in the voltage regulator and rectifier, as well as the transient voltage spikes, the harmonics from the mains frequency and anything else picked up by the mains cables on their journey from the power station.

Householders, who are connected for electricity and the telephone, understand that they receive these services through the company's cable. What is less easy to understand is that their connection to the cable allows them not only to receive but also to transmit disturbances to the telephone or power company, an unfortunate fact which must be assimilated.

Mains power is 50 Hz at 230 volts $\pm 6\%$ but every time the demand changes because someone switches an appliance on or off, alterations are forced on to the voltage. Spikes of double the nominal supply voltage, up to 500 volts (V), are liable to happen every day. Spikes between 1000 and 6000 V are likely once a year. The daily spike of double the nominal voltage is intolerable to many users and there are many ways of keeping it out of their supply.

In addition to switching spikes, there is other interference. A mains cable is long enough to pick up any interference near it by acting as an antenna. But mains supplies at points far from the generator also suffer voltage drop under heavy load when much current is being taken. This is usually within the allowable 6% but, in industrial premises, heavy loads from industrial machines may bring a further voltage drop, totalling up to 10%.

A further source of interference is mains signalling between 3 kHz and 148.5 kHz, according to EN 50065-1:1992. It may have to be removed by connecting a filter to the power supply. Because mains-signalling can save large sums of money by dispensing with the need for meter-readers to call at people's homes, its use is certain to expand. There are also other uses of mains signalling, such as for intercoms and switching.

Fuses and circuit breakers, which are essential for safety reasons, protect electrical installations from overloads but they are not helpful to EMC because they cut off the power abruptly, and create interference. They are also too slow to be useful in reducing transients.

Power generation

To provide a truly smooth voltage there are either mechanical or electrical methods. Mechanical methods mean generating one's own power, which involves the need to employ someone who understands electrical generation and supply, or the easier alternative that is feasible in western Europe – a motor–generator set (MG set). However, for electrical consumers who lack the investment needed for the massive concrete foundations and heavy machines of an MG set or who have no space to install one, there are many other solutions to mains voltage problems.

Motor generators

Providing electrical power by one's own generator, whether by diesel or petrol engine, is unusual in an industrialized country but may be unavoidable where the mains power supply is less certain.

For regular consumers of large amounts of power, the considerable investment in an MG set can be worth while since an MG set is less trouble, highly reliable, efficient and practical where adequate and reliable mains power exists. An MG set is an electric motor driving an electric generator. Usually both work at the same speed, so they can be on the same shaft, but for EMC purposes a break in the continuity of the drive shaft is most helpful in keeping the output voltage smooth. It interrupts any stray, disturbing current. A universal joint or similar device in the drive shaft can effectively make this break without mechanical difficulty.

Electric motors and generators are extremely reliable and work without interruption for years with very little maintenance. Both have high efficiency, usually well above 95%, so the total power loss in the MG set can be expected to be well below 10%, giving an efficiency of better than 90%. Electric motors are very tolerant of spikes, noise and short interruptions of supply. Because of the flywheel effect of the two machines together, even a complete interruption of power for several seconds may pass almost unnoticed when the power is provided by an MG set. If the motor in the MG set is brushless it can be expected to make no commutator noise, which is always a nuisance with commutator motors.

Electrical methods

Electrical methods may be cheaper than mechanical and have no disadvantages of vibration and noise. Several types of voltage regulator have been known to and used by AC power supply companies for many years. Six ways of ensuring a smooth voltage in the power supply, although not by voltage regulation, are mentioned below, namely, filters, optical links, isolating transformers, ferroresonant transformers, power conditioners and uninterruptible power supplies.

Probably the simplest method is to connect one or more filter networks to the mains supply. Power-line filters may be bought complete, combining a transient suppressor with the filter. Filters need very little space and are discussed at greater length in Chapter 11.

An optical communication link using glass or plastic fibre is immune from electrical interference. The optical receiver as well as the transmitter are electrically operated so they should be carefully installed with a 'clean' power supply. Unwanted interference can be misguidedly transferred to the optical side if the transfer between electricity and optics is not meticulous. An opto-isolator can isolate two signal circuits from each other even at voltage differences as high as 4000 V. It is also more efficient than an isolating transformer but is probably much better suited to the small currents used in signalling than to the large currents of power supplies.

Isolating transformers, usually well shielded, are used in all low-noise equipment. Instrument transformers have protective shields between the windings to reduce or eliminate interference. For life-supporting medical equipment these are essential and it is engineering malpractice to install an instrument without a well-shielded isolating transformer. The so-called inter-winding shields may be copper foil interleaved between the windings or a complete metal box around the transformer, or both together.

A ferroresonant transformer is a transformer designed to work with its magnetic core saturated, unlike the usual situation – avoiding saturation. Operating at a high primary current, the saturated core will not greatly increase the output even with a sharp voltage spike. An ordinary transformer, if saturated, can thus be used as a voltage regulator, even if the input primary current is almost as high as a short-circuit current.

A power conditioner generally costs less than an uninterruptible power supply (UPS) because it does not have the expensive battery, but uses electronic circuitry, such as an amplifier with a transformer. This cancels spikes, under-voltages, oscillations and power outages of short duration. But power conditioners have the disadvantage that power outages of more than a few milliseconds can bring a complete shutdown.

Uninterruptible power supplies rely on a battery which is trickle charged through a rectifier from the mains supply. The battery cells need space and are heavy but they do not make the noise and vibration which are unavoidable with an MG set. The battery power is converted to the AC supply voltage by a solid-state inverter (static inverter). The best and most expensive UPSs have a diesel- or petrol-driven generator set that starts up automatically after some minutes of power stoppage.

Power factor improvement

Power factor improvement has been used by electricity distribution companies for many years. It helps to improve the smoothness of the voltage and to reduce the power losses from the power supply. Unfortunately any switching in or out of capacitors or inductors used for power factor improvement causes interference. (See Glossary for an explanation of power factor, and Fig. 11.2, Chapter 11, Filters, for symbols.)

Common-mode choke

Interference in the form of high-frequency noise that is brought by the mains can be excluded by a relatively simple device, although it may be laborious for people unfamiliar with electrical work. It involves connecting a common-mode choke to the power-supply cable, as near as possible to the radio or TV set or other device that is suffering interference.

A common-mode choke is usually wound on a ferrite ring, although a rod may help if a ring is too awkward to use. The rings may be called beads. The difficulty with a ferrite ring is that the power supply cable should be wound into it several times, which is not easy to do unless the cable is very thin. This involves temporarily removing the plug from the cable for insertion through the ferrite ring.

One way of avoiding such difficulties is to reserve a special short and thin cable for the purpose, which can be inserted on any device that suffers mains interference. The thin cable with a ferrite ring or rings on it has a socket at one end and a plug at the other, if necessary, that is suitable for connecting coaxial cable. This is an easy, quick way of inserting a ferrite ring on to any cable. This special short, thin cable with connectors for coaxial cable at each end can be inserted quickly into a length of coaxial cable with no trouble. The correct way of winding the cable on to the ring is shown in Fig. 3.1.

Usually four turns of cable on a ring should be enough although a specialist opinion, both on the type of ferrite and the number of turns, is always advisable. If the cable is too thick to pass more than once through the ferrite, it may be necessary to buy several ferrite rings so as to pass the cable once through them all.

Old (and new) portable radio sets usually have an internal antenna made of a ferrite rod. When the radio is scrapped it is worth extracting the ferrite rod for future use because ferrite rings are expensive (in 1996, between £1 and £2 per ring). The special short cable mentioned

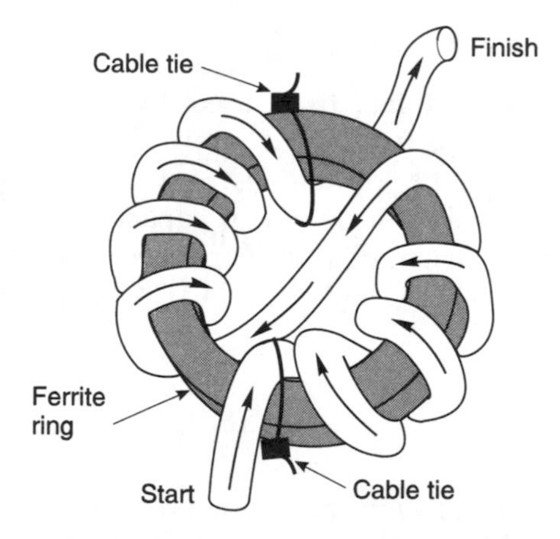

Fig. 3.1 Layout of a common-mode choke made from a ferrite ring with power supply cable correctly wound round the ring (source: Radio Society of Great Britain)

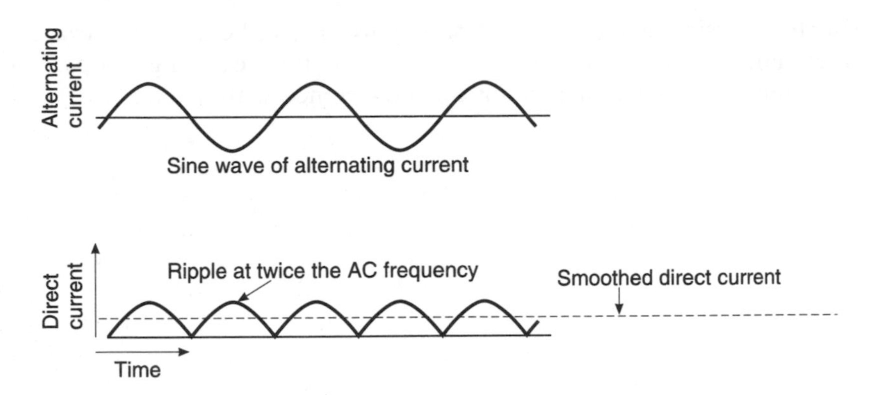

Fig. 3.2 Rectification of AC sine wave (above) showing (below) DC ripple and DC after smoothing

can be made with a ferrite rod but usually the cable has to be wrapped with more turns than are needed for the ring, often nearly 20.

The effectiveness of the device depends on the square of the number of turns multiplied by the number of rings. Thus 14 turns in two rings is equivalent to:

$$14^2 \times 2 = 196 \times 2 = 392$$

With only seven turns, equal effectiveness could be achieved only by buying six more rings:

$$7^2 \times 8 = 49 \times 8 = 392$$

But as usual, professional advice is helpful.

Computers

Computers are normally digital systems on low-voltage DC but operating at radio frequency, usually a high frequency of several MHz. They are capable both of radiating interference and of being victims of it if their circuits or cables pick it up. DC power for a computer, if obtained directly from AC mains, is put through a rectifier that leaves ripple at twice the mains frequency. The DC ripple (Fig. 3.2) is removed electronically together with any noise generated in the semiconductor devices or brought in by the mains.

Some operational amplifiers (opamps) are described as 'low-noise'. At low frequencies these can have excellent performance, rejecting 100 dB of noise from the power supply.

Other amplifiers, however, can be less satisfactory. The power supplies needed for the biasing of sensors should also be 'clean'. The shortest possible length of connecting wire should be used between any filter and an amplifier or sensor, eliminating earthing loops and reducing any chance of picking up interference with a long wire.

Chapter 4
Radio waves and how they travel (propagation)

Interference travels along the same routes as wanted radio waves. Radio waves use three different ways of travelling, two of them involving reflection either from the ground or the ionosphere or both. The atmosphere is the third medium. The most stable transmissions take place near the ground because the ionosphere especially suffers continuous fluctuations. We are all familiar with weather disturbances, but the ionosphere is even more disturbed, reversing its condition every 12 hours. An outline of radio-wave travel is therefore given in this chapter. Radio receivers need careful design if they are to provide intelligible information from very weak signals in the presence of heavy interference.

Three common methods of travel for radio waves include two types of direct wave wholly through the atmosphere. One is a straight line from transmitter to receiver. The other, because of refraction, curves gently down towards the earth. It can therefore travel beyond the seen horizon. The third type is reflected from the ground.

The two commonest routes of radio waves are along the ground (surface waves, which often do not travel far), and the sky-wave methods which can cover thousands of kilometres, using reflection from one or other layer in the ionosphere, sometimes including re-reflection from the earth. The earliest radio involved ground waves through the lower part of the atmosphere, normally called long or medium wave.

Radio listeners know how easy it becomes every day after nightfall to listen for hours to distant stations that are inaccessible by day (Fig. 4.1). Then, at dawn, when the sun comes up, the distant stations again

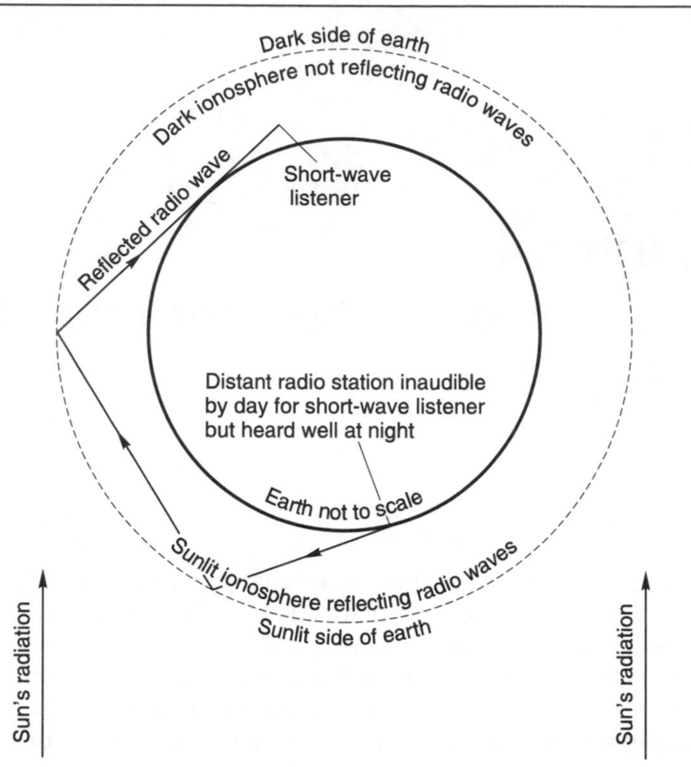

Fig. 4.1 Distant radio stations become audible after nightfall (not to scale)

escape. The ionosphere, a spherical collection of layers roughly 50–400 km above the earth, is reloaded by the sun daily with ultraviolet radiation and electrically-charged particles including electrons and ions, hence its name. With nightfall the sun ceases to supply ions and the ionosphere, becoming dark, reverts to a more inert state.

The explanation of this is that broadcasts are reflected nightly to the dark side of the earth from its sunlit side where the ionosphere is most active because of being recharged by the sun.

Weather forecasts are difficult and unreliable. Forecasts of the ionosphere's behaviour are even more difficult because it is so far away (50–400 km) and much of the evidence for forecasting what will happen in it comes from watching the sun, its sunspots and flares millions of kilometers away. Nevertheless, telecommunications organizations can make their essential forecasts and thereby achieve good communications.

Radio waves can be received over long distances as sky waves, skipping over the earth's surface by reflection from the ionosphere (Fig. 4.2). Double or multi-hops also exist, in which the radio waves

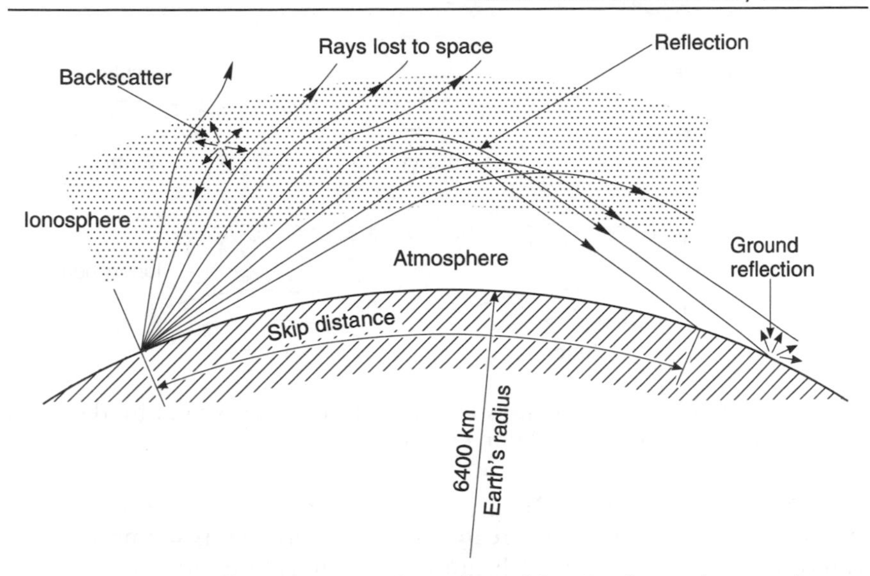

Fig. 4.2 How radio waves may be reflected by the ionosphere or lost to space (after Hall and Barclay, 1989)

bounce up from the earth's surface or from another layer of the ionosphere one or more times (Fig. 4.3). The longest single hop is 4000 km after which more than one hop is needed. In this connection one must remember that seawater gives the best reflection for radio waves and dry rock or sand gives the worst. Each time an RF wave bounces up from the earth's surface it loses strength, usually 4 dBs.

A summary description of the ionosphere and its solar origins is given here.

The sun

The sun, a vast nuclear furnace nearly 150 million kilometres from the earth and about 108 times its diameter, provides the heat needed for all life on earth as well as what is needed for our weather and its changes. This is perhaps less surprising when one realizes that the sun contains 99% of the mass of the solar system. Solar activity also affects radio interference. The sun's rays at ground level on the equator provide an average 1.1 kW of heat for every square metre of receiving surface, a stupendous amount, as we realize when out in the sun on a clear midsummer day. More than this amount of heat reaches the higher levels in the ionosphere or upper atmosphere.

But the sun also sends us the solar wind, a stream of ultraviolet and other radiation, electrons and other particles. This stream does not reach

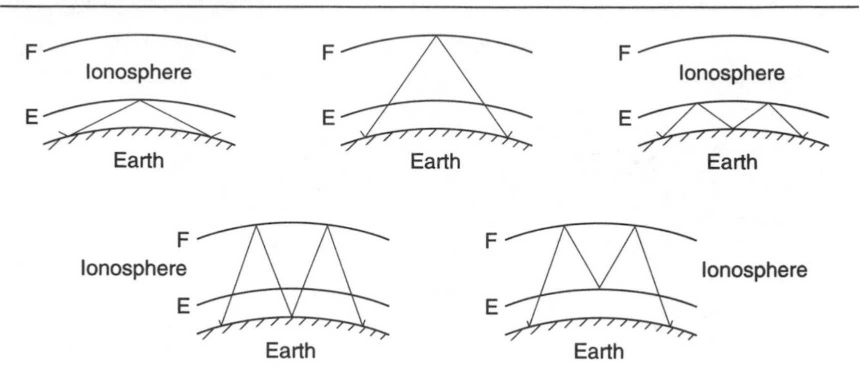

Fig. 4.3 Possible different ways of reflection of radio waves by the E- or F-layers of the ionosphere (after Picquenard, 1974)

the solid earth, only the upper atmosphere where its EM waves and electrically-charged particles create the ionosphere that is so important to radio propagation over long distances, as well as to interference.

The sun is made of hot gases, which at its visible surface are chiefly hydrogen and gaseous calcium, unlike the cool earth which is largely solid or liquid. Like the earth, the sun rotates, but being gaseous, its poles turn more slowly than its equator. The poles turn completely once in 35 days, the equator once in 27 days. This twisting of the sun's surface disturbs it, creating storms, flares and sunspots.

Sunspots are areas on the sun's surface, known since antiquity, which, although not dark, seem dark because they emit less light than most of the sun's surface. This is why in antiquity some writers believed that sunspots cooled the sun. But the truth is probably the opposite, that sunspots make us warmer. They are whirlpools of hot gases which are strong electrical generators, and because of this they are also the centres of intense magnetic fields. Yet another magnetic effect from the sun is caused by the electrons and other charged particles brought to the upper layers of our atmosphere from flares and sunspots by the solar wind. They are deflected around the outer atmosphere, creating a ring current round the earth that affects the earth's magnetic field, the magnetosphere, some 100 km up.

Mini Ice Age

So our magnetic environment is strongly affected by the sun. Variations in the number of sunspots would therefore be of interest if only for this reason, but sunspots also seem to have an effect on our weather. The mini Ice Age, 1645–1712, apart from involving the greatest advance of mountain glaciers in modern times, coincided with the rise of the

Puritans in England and was a time of sunspot minimum, accompanied by poor harvests. The Scots are believed to have been persuaded to agree to the Act of Union of 1706 that united England and Scotland only because in parts of Scotland famine seemed near.

Past sunspot cycles have been established by measurements of carbon14 (also written 14C) made on the wood of old trees from centuries-old forests. From these measurements it has been deduced that repetitions of sunspot maxima or minima occur in cycles. For many years sunspot cycles were believed to last only 11 years, but astronomers now believe 22 years is more likely. The most recent sunspot maxima occurred in 1947, 1958, 1969, 1980 and 1991. The present series of sunspot cycles began in March 1755 with Cycle 1. Cycle 23 begins in 1997. The succession of 11-year sunspot cycles has been worked back to 527 BC, many of them confirmed by the thicknesses of clay laid down in large lakes.

But it is only fair to point out that many other, much longer climatic cycles have been devised by climatologists. The wandering lake, Lop Nor, in Sinkiang, central Asia, alternates between a northern and a southern location for 1700 years each, totalling 3400 years. The Nile floods have been estimated to have cycles of 84, 168 and 737 years by different scientists. Another climatic effect is that of major volcanic eruptions. The dust they throw up can cool the world for several years by cutting off sunlight.

Smith and Best (1989), quoting Chizhevsky and others, point out that 80% of historical events, such as revolutions, rebellions or outbreaks of war, have occurred during periods of maximum or rising sunspot activity, often simultaneously with magnetic storms, and only 5% during sunspot minima. This applies to the French Revolutions of 1789, 1830 and 1848, the Commune of 1870, the Russian revolt of 1905 and the 1917 revolution, the outbreak of war in 1939, the invasion of Czechoslovakia in 1968, the invasion of Afghanistan in 1979 and the Falklands war in 1982. With hindsight we can now add the events at the end of 1989 in East Germany and Romania.

In mid-March 1989 (*New Scientist*, 3 Feb. 1996) an exceptionally fierce magnetic storm in Quebec extinguished the lights of 6000 people who depended on the power grid. The same storm brought the northern lights to places well south of London. The next such maximum is expected around 2000 and 2001 when we can foresee similar disastrous events.

Ionosphere: the D-layer

The ionosphere includes several continually changing layers of plasma (ionized gas) more than 50 km above the earth, all of which can assist radio propagation by reflecting radio waves back to the earth. They can also prevent radio propagation by absorbing radiation or by letting it pass through to outer space. The lowest layer, the D-layer, is from 55–95 km up and reflects low-frequency radio waves at frequencies probably below 300 kHz, but is also regarded as a considerable absorber. It fades away every night when distant radio stations become audible, but as it reappears with the dawn, the distant radio stations again escape. In winter in northern latitudes the D-layer disappears.

The ionosphere affects the propagation of all waves at least up to 50 MHz. Frequencies below about 30 MHz are usually reflected. Those between 30 and 50 MHz can be reflected by scatter. The ionosphere can be imagined as a series of thin layers. A radio wave from earth striking the first layer may be partly reflected but another part of it is lost to space (transmitted). The transmitted part may be reflected back towards earth at a higher layer and may then reach earth or be reflected away. There are many possibilities.

E-Layer and sporadic 'E'

The next layer up, the E-layer, was the first one discovered in the early years of this century. It was then known as the Appleton or Kennelly–Heaviside layer from three of the US and British scientists who helped to establish its presence. It is above the D-layer, from 90–125 km up. Because the ionization comes from the sun, all layers are most strongly ionized at noon, and some disappear at night. Naturally, they are more strongly ionized in summer than in winter.

Sporadic-E layers are irregular ionized layers about 100 km up, often only 1 km thick and 70–150 square kilometres in area, which may appear during auroral displays, unpredictably, but often in summer. Their origin is a mystery but they can be useful for short-lived communications because they reflect RF waves above 30 MHz which are not reflected by other parts of the ionosphere. They often become continuous near the magnetic poles. In the days when television was broadcast on VHF (405-line TV) it was not unheard of to receive New York taxicabs on British TV sets.

F-Layers

Still higher is the F-layer or layers, several hundred kilometres up, with the highest ionization of all. Although some ionization is present even at night, most occurs in the daytime, when it splits into two layers, F1 and F2. The F2, the higher, extends in summer up to 500 km above the earth. These high reflecting layers provide the most long-distance propagation on frequencies from 2–30 MHz. Normally they do not reflect frequencies above 30 MHz and sometimes the limit is even lower although occasionally frequencies above 50 MHz have been reflected. It is normally only by day on the sunlit hemisphere that ionospheric reflection can help radio propagation.

Figure 4.3 shows how much further a wave can go if it is reflected from the F-layer rather than from the E-layer. A single hop from the F-layer can travel only as far as a double hop from the E-layer. Such decisions are unfortunately not under the control of telecommunications authorities, they are decided by the ionosphere.

Skip zones are areas of the earth between transmitter and receiver that cannot be reached either by ground wave or by ionosphere reflection. They are too far to be reached by ground wave and the sky wave passes overhead.

Sudden ionospheric disturbance

All the layers suffer daily variations as well as seasonal, geographical and other changes caused by the 11- or 22-year sunspot cycles. Other disturbances can affect the ionosphere, and these also originate in the sun. The main ones are solar flares which can be seen on earth by people who observe the sun. Solar flares are an intense brightening of a small area, usually close to a sunspot. There are two main emissions from solar flares, first their EM radiation travelling at the speed of light, and secondly their ejections of highly charged particles, travelling more slowly. The EM radiation arrives in some 500 seconds since the sun is 150 million miles away, but the particles take much longer, usually arriving 18–36 hours after the flare.

The early arrivals, UV emission, gamma and X-rays can cause a radio blackout and an increase in electronic noise at frequencies below 300 MHz. They can also cause a sudden ionospheric disturbance (SID). The ionization of all layers rises abruptly, even in the D-layer. The lower frequencies are the first to be affected by a SID and are the last to return to normal. Just the sunlit side of the earth is affected, as would be expected. If the dark side of the earth can be used for communications, the effects of a SID are often avoided. Occasionally it may be possible to communicate on the sunlit side of the earth using the higher frequencies.

After the flare, usually when the SID has died away, the electrically charged particles ejected from the flare arrive at the earth causing storms in the ionosphere and radio blackouts that can last for a day or more. On reaching the polar regions they cause the aurora borealis in the north and the southern lights near the south pole. During storms in the ionosphere the lower frequencies are more likely to provide communications especially on paths that cross the equator. If the F-layer completely disappears, as it can in such storms, any ability to communicate on higher frequencies is also lost. The SIDs happen first, at the same time as the radio blackouts, 500 seconds after the solar flares, these are followed by the storms and auroras.

The auroras in the north or south polar regions can help radio communications by their reflections of radio waves directed at them, as radio amateurs have learned.

Sunspots grow and die but they may last for months. Because the sun rotates once in 27 days, the previous sunspots of 27 days earlier often reappear and recreate the previous month's conditions in the earth's ionosphere. Sometimes the same applies to flares although they are usually more short-lived.

The ionosphere layers are far from being invariably reliable. They sometimes reflect radio waves and help radio propagation. At other times and in other places there is no reflection but only absorption or complete disappearance into outer space of any radio waves sent up.

Critical frequency

The altitudes of the different layers of the ionosphere were found by experimental radar pulses sent up steeply. From pulses sent at various frequencies it was found that different 'critical frequencies' existed everywhere, above which no pulses returned to earth. Layers with the highest critical frequency are farthest from the ground. The critical frequency is rarely higher than 50 MHz and at night in the E-layer varies from 0.25–0.5 MHz. The critical frequency of the lowest layer, the D-layer, is only from 0.1–0.7 MHz. The critical frequency varies everywhere not only hourly but also seasonally and geographically. In 1927 during an eclipse of the sun the critical frequency diminished abruptly as the eclipse progressed. The critical frequency at any place is closely related to the MUF (maximum usable frequency) of any transmitter nearby.

Experiments have shown that the main agent creating the ionosphere is ultraviolet emission (UV) from the sun. The very low gas pressure and variable gas densities in the ionosphere at great heights of hundreds of kilometres allow the UV to generate many free electrons there.

Multipath effects and digital audio broadcasting

An explanation is needed for the annoying multipath disturbances which occur in all modes of reception. If an EM wave can reach a receiver by two or more paths, because of reflections off buildings or hills, there is a possibility that a wave arriving along a short path will destructively interfere with a wave arriving by a long path, because it arrives first. If one arrives exactly half a wavelength before or after the other, the two will cancel each other completely. There will be no disturbance because nothing will be received. But this is uncommon and usually the reception is disturbed. One can, however, now hope that multipath disturbance will soon be a thing of the past, at least for radio in London, because the BBC announced the introduction of DAB (Digital Audio Broadcasting) into the London area on some of its main radio programmes in September 1995. DAB is said to eliminate multipath effects, but in early 1996 only about 30 DAB receivers were in use, worth some £600 each.

Different paths come into existence because an EM wave can be reflected from a hillside or from a building or other obstacle. Multipath disturbance (Fig. 4.4) can sometimes be avoided by placing the antenna in such a way that an unwanted path is masked by a chimney, building or other obstacle so that it cannot reach the antenna. A directional antenna aimed at the wanted path can also be effective in excluding an unwanted wave.

London's tallest skyscraper at Canary Wharf in the docks has the unfortunate distinction of causing multipath disturbance by reflections from its smooth metallic wall cladding and metallized glass windows. This is in addition to the fact that it prevents any reception by those beyond it who are in the direct line of sight from the transmitter. Unfortunately the radio and TV authorities have been unable to help.

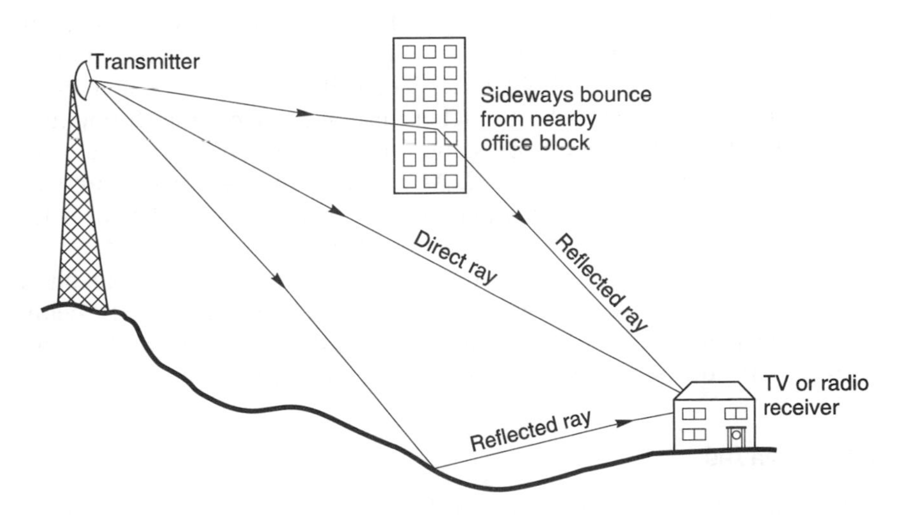

Fig. 4.4 Multipath journeys of radio wave to receiver from transmitter

Ground waves (surface waves)

Even if the target of a transmission is beyond the horizon at the transmitter, it may be possible for longer wavelengths to reach it since, by refraction or diffraction, they bend round to follow the earth's curvature and avoid obstacles. But there are two sources of interference, either a ray reflected from the ground or one reflected from the ionosphere. Reflection from the ionosphere is most likely at frequencies of 1.5–30 MHz (HF or short wave). Either of these reflected waves, because of multipath effects, can cancel or disturb the desired signal. There may be additional disturbances from the sky wave caused by fluctuations in the ionosphere.

Refraction

The temperature of the atmosphere normally drops 1°C per 150 m increase in height. This causes a change in the refractive index of the air, which helps the EM waves to bend round the earth and follow its curvature. This effect is important for radiation travelling at small angles of elevation but does not affect waves aimed steeply upwards.

Ground-wave transmission at frequencies from 0.1 to 3.0 MHz (wavelengths of 3 km to 100 m) can cover long distances over water and shorter distances over land. The navigation systems of LORAN use these frequencies.

Troposphere and tropopause

The troposphere, the lowest layer of the atmosphere, averages about 11 km deep, ranging from 8 km at the poles to about 16 km at the equator, and is crowned by the tropopause. Throughout the troposphere the temperature drops as one climbs higher until one reaches the tropopause. But at the tropopause and for some 20 km above it the temperature remains constant at about −55°C. At this level the barometric pressure has fallen to some 7% of its value at the earth's surface. Above it the temperature rises again but the pressure continues to drop.

The troposphere contains all the water and dust of the atmosphere and 75% of its mass. The water and dust, at altitudes mainly below 1 km, are helpful to radio propagation because they allow 'bending' (refraction) and 'scattering' that is known as troposcatter.

Microwaves

Microwave propagation is one of the many established methods of telephoning as well as of data transmission. It uses repeater towers visible to each other, perhaps 30 km apart, consequently employing LOS (line of sight) communication. The reliability of microwave communications is excellent, as indicated by the fact that many railways now use the system. They began to replace the old wires of the railway telegraphs from 1959 onwards. However, those who like modern methods such as satellites or meteor burst communications believe microwave communications may be obsolescent.

Although ordinary electronic units like capacitors, integrated circuits, resistors, transistors, even ordinary wiring and printed circuit boards do not work well at microwave frequencies, and special devices have had to be invented, it is worth dwelling briefly on the extraordinary abilities of microwave, apart from cooking at 2.45 GHz. These abilities, related to its cooking ability, concern power transmission.

Stratospheric relay platforms

Several projects have appeared for unpiloted aircraft to circle at about 20 km up and supersede the artificial satellites now in use. The stratosphere is attractive because it is relatively devoid of atmospheric storms. The aircraft would use microwave power beamed up from the ground, but so far none of them seem to have been put put into practice. A quite different project in Japan, possibly more practical, the HALROP (High Altitude Long Range Observation Platform) is being undertaken by Dr M. Onda, Chief Engineer of the Government Mechanical Engineering Laboratory at Tsukuba (Fig. 4.5). He proposes a 200 m long unmanned remote-controlled helium-filled airship, also at 20 km altitude. Several models about 20 m long have been flown successfully.

Near the equator in Japan solar panels would provide propulsion energy but farther from the equator microwave power would be beamed up from the ground. This unmanned, non-rigid, mainly stationary airship would, like an artificial satellite, relay telecommunications, monitor the environment or the weather, coordinate disaster relief and in polar regions possibly monitor or repair the ozone layer.

The main aim of this research is to develop, in a few years' time, working airships to be built by Japan's distressed shipbuilding industry. The ballast will be atmospheric air contained in spherical 'ballonets' in the helium-filled spaces. Air can be released to outside whenever the airship needs to climb higher. At 20 km altitude, atmospheric air has only 7% of its density at ground level.

In autumn 1995, a test airship 16.5 m long stayed 50 m up under remote control for several minutes powered by microwaves at

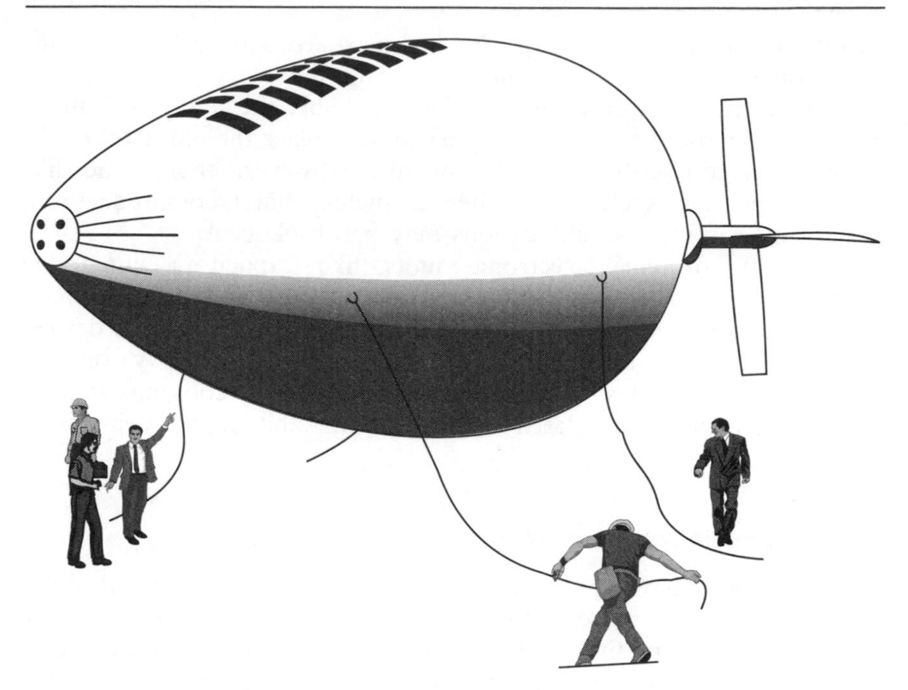

Fig. 4.5 The 20 m long, non-rigid solar-powered HALROP in 1992, seconds after take-off, showing its solar cells on the upper surface and on the right its inflated cruciform steering tail (empennage)

2.45 GHz beamed up from the ground. Its rectifying antennas measured about 3 × 3 m, weighed about 30 kg and sent 5 kW of total power to the two electric motors of 2.5 kW each which drove two thrusters.

Current forecasts for the 200 m long HALROP, with a pay load of 3 tonnes, are that it will weigh 24.5 tonnes of which one-third will come from solar cells and batteries and another third from the envelope. The 400 000 cubic metres of helium will easily lift this load. The main motor of 100 kW will weigh 500 kg.

Temporary modes of propagation

Many temporary but usable modes of propagation of radio waves exist, some of which can be counted on only for a few seconds. These modes include ducting, grey-line propagation, lamellae, equatorial (trans-equatorial) propagation, and 'scatter' modes, from the ionosphere, or meteorites, or the troposphere (troposcatter).

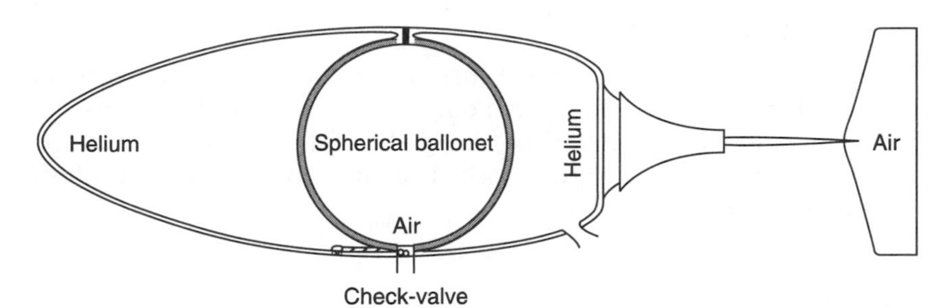

Fig. 4.6 Cross section through model non-rigid helium-filled HALROP (high-altitude long-range observation platform) designed and built by the Japanese Mechanical Engineering Laboratory at Tsukuba, Japan, showing the gas-filled steering empennage at stern

Ducting and music from whales

Ducting of radio waves for long distances over the earth can happen when a mass of warm air settles over cooler air at the ground surface. This 'temperature inversion' is so called because normally the air gets cooler as one climbs, but an inversion reverses this trend, the warmer air is at the top. The boundary between warm and cool air, the edge of the duct, acts as a reflector for the radio waves, keeping them within the inversion. A duct need not be close to the ground, although it usually is. It can be an 'elevated duct' well above ground and caused by a similar temperature difference with warm air above cold air. Ducts appear only in quiet weather and above water or plains. They can be as little as 6 or 7 m thick but they may reach to 160 m. Below 50 MHz, ducting is rare and it has not been noticed at frequencies above 450 MHz.

Evaporation ducts are probably the most useful and commonest of the many forms of duct. They exist over any water surface of reasonable size, especially the sea. The decrease of water vapour in the air from water level to some metres higher is the main reason for it. An evaporation duct averages 3–4 m thick along the northern Norwegian coast, 6 m along the west coast of Norway, 5 m in the North Sea but 6 m between England and France and from 10–13 m in the Mediterranean.

Although sound is transmitted quite differently from radio waves it can also be sent thousands of miles over the ocean surface, along ducts possibly caused by temperature inversions. In his book *Lifetide* (1979), p. 127, the biologist Lyall Watson describes the song of the hump-backed whale, *Megaptera novaeangliae*, off Bermuda, pointing out that its approximately 200 notes can be picked up by another whale many miles away who returns the tune with musical embellishments.

Grey-line propagation

The 'grey line' in this sense is the line across the globe between night and day. This lasts for about an hour when the D- and E-layers are reappearing in the sunlit half of the globe. Efficient communication between distant corners of the world becomes possible for this short time, around 6 a.m.

Lamellae

Lamellae (small layers) seem to be permanently present in the troposphere. They can provide weak reflections. They are at altitudes from 300 m to 3 km, can be up to 20 m thick, but are no more than a few kilometres in horizontal extent.

Trans-equatorial propagation

Equatorial transmission by the ionosphere was first noticed in 1947. It allows transmission at 50 MHz or up to 1.5 times the predicted maximum usable frequency. Year round, the ionosphere is more stable in the tropics because it has more sunlight than the temperate regions and its critical frequencies there are always high. Trans-equatorial (TE) mode is believed to result from masses of ionized particles in strips only about 10 m thick, up to about 1000 km long from east to west but about 1 km wide from north to south. TE mode normally occurs on autumn nights.

Scatter modes

Clouds of ionized particles in the ionosphere allow reflection by scatter over a wide number of frequencies. It is not known if these are strongly ionized clouds or more turbulent regions of the ionosphere. They are, in any case, not part of its recognized layers. Ionospheric reflection is widely preferred for long-range propagation at frequencies below 30 MHz. But ionospheric scatter also can cover long distances, from 900 to 2000 km, at frequencies of 30–100 MHz.

Secondly, troposcatter, with a high-powered transmitter, using reflections from dust or water particles in the atmosphere, can cover distances of several hundred kilometres, on a frequency range from about 500 MHz up to 5 GHz.

Thirdly, meteorite scatter is based on reflections from the ionized trails (shooting stars) left by meteorites as they burn up while they traverse the rarefied air in the ionosphere. All meteorites begin to burn as they enter the E-layer because of frictional heat, leaving an ionized trail behind them. Meteorite scatter can be used for communications of 2000 km distance at frequencies of 50 MHz or higher, and for a few hundred kilometres at 144 MHz.

Meteor-burst communication

One special variety of meteor scatter is meteor-burst communication (MBC), which became popular in the USA only in the late 1970s when its essential automatic computer control became easily available. It provides 24-hourly communication over a radius of 2000 km using inexpensive battery-powered terminals with small antennas which access the trails of some of the many millions of tiny meteorites about 100 km up, often for only a few seconds at a time. It is unsuitable for speech because a short wait is usually needed to find a suitable ionized trail to reflect the signal, but it is appropriate to the sort of message that was formerly sent by teletype, usually not needing an immediate reply. It is unlikely to suffer interference.

Data rates during transmission bursts vary from 2000 to 9600 bits per second (about 30–150 words per second) but because of the variability of the trails, the average speed is about one-tenth of these levels. Operating frequencies are between 30 and 50 MHz.

Another reason for the popularity of MBC is that it is not vulnerable to nuclear attack. The footprint of the signal is about 5 km wide by 25 km long. This tiny fraction of the 4000 km diameter implies that jamming is unlikely, and so is unintentional interference. Short-range communications are possible for any distance up to 2000 km but for very small distances line-of-sight communication is more likely. The Alaskan Air Command has 13 high-power (10 kW) meteor-burst terminals which are a cost-effective backup to the normal satellite link. Satellites can be destroyed. But the civilian Alaskan MBC provides services for US engineers, weather and soil conservation. 'Hydro-meteor' scatter, reflected from rain, snow, hail or ice clouds, is important only at SHF (super high frequencies) from about 8–20 GHz.

Mobile radio

Mobile radio, being usually in or near a city, often cannot use a line-of-sight path and so it has to succeed by scatter or by multiple reflections from obstacles, mainly buildings. The movement of a vehicle carrying the radio receiver can complicate reception by adding a Doppler shift in frequency that varies with the car speed.

Moonbounce

A review of propagation would be incomplete without a mention of moonbounce (EME or earth–moon–earth), even though it is far from being commercial. It was first achieved in May 1946 by an ex-radio amateur, John De Witt, then a lieutenant colonel in charge of the US Army's Evans Signal Lab in New Jersey, USA, who was trying to answer the US government's question: 'Could an enemy direct a radio-controlled rocket at the USA?' De Witt sent out one-second radar pulses at 144 MHz frequency directed at the moon at 4-second intervals and succeeded in receiving echoes of them.

Doppler shift

A Doppler shift is the change in frequency given to a wave by the movement either of the body that transmits the wave, or of the body that receives it or both. Most people have noticed the Doppler shift in the change of sound of the railway engine hooter as a train passes by. But those who have to calculate the Doppler shift in moonbounce are sometimes confused by a '2' in the Doppler formula. The '2' is there because there are two almost simultaneous Doppler shifts, one for the transmitted wave and a second for the reflected wave.

The moon's distance from earth varies between 356 000 and 406 000 km, which reveals one of the complications of moonbounce. Some of the time the moon is approaching the earth, at other times it is receding, so there are opposite Doppler shifts of the frequency for the two directions of movement, corresponding to the moon's approach- or recede-velocity of up to 1500 km/h. Another complication of EME is that the moon wobbles as it rotates, causing the reception to fade. EME is not and never was a commercial proposition, merely a fascinating demonstration for radio amateurs of what can be achieved with highly sophisticated and very expensive equipment allied to great perseverance both at the transmitter and at the receiver.

Chapter 5
Antennas

Antennas (aerials) are usually made of metal wire, tube or rod and send out or receive electromagnetic waves. They come in many shapes and sizes depending on such variables as the frequency of the radio wave, its strength and its direction of sending or receiving. The wide variety of antennas now available includes monopoles (wires in attics or hung from windows) as well as the more complex horns, dishes (parabolic reflectors) and phased arrays. In this little book there is space to describe only two of them, the simple dipole and the more complex beam or directional antenna. All antennas are compared with a theoretically perfect omnidirectional antenna, able to transmit in all directions with the same strength. An antenna is made more directional by concentrating its power in a relatively small area whether for transmitting or receiving. A beam antenna wastes less energy and so it is more efficient in using the power available to it than one that is omnidirectional.

In the EMC testing of appliances, antennas are the key components, and even electrical supply cables can act as antennas. What the antenna receives and how the surroundings affect the measurements are usually unknown, especially for tests made in the open where there is usually too much ambient noise. Antenna theory is complex and highly mathematical so cannot be treated in this book. Antennas for EMC work must be cheap and rugged, but few can cover enough frequencies. Different antennas may have to be used – each specialized for its own band of frequencies.

The four main factors affecting an antenna's ability to concentrate its power and reduce unwanted radio waves are bandwidth, gain, beamwidth and polarization.

Bandwidth

Bandwidth is the range of frequencies around the central frequency that an antenna can receive or transmit. An antenna is usually built to work on a specific range of frequencies around some central frequency. Within this 'band' such properties as gain, beamwidth and polarization are as close as possible to that at the desired central frequency.

Gain

'Gain' tells us how much more power a directional antenna has in one direction than an ideal omnidirectional (isotropic) antenna. Antenna gain can be compared with the light that would come from a torch bulb alone, as opposed to the much more powerful, focused beam one gets from a torch with its reflecting mirrors in place.

Beamwidth

This brings us to another variable of special interest to antennas, called beamwidth (or 3dB beamwidth). It is the width angle in a stated plane across the direction of radiation of a directional antenna. It is usually defined as the angle between points either side of the beam centre line at which the power has dropped to half (3 dB less than) that on the centre line. In our torchlight analogy, if one was to shine a torch at a movie screen, then the beamwidth would be the angle at the torch bulb between two points, A and B, possessing a light intensity equal to half that at the centre of the light beam (Fig. 5.1).

Polarization

Another important variable in ensuring that the strongest possible radio wave is received by the antenna is to ensure that the antenna's polarization matches that of the incoming radio wave. Polarization is, by definition, the angle made by the plane of the electric wave with the horizontal (the ground). The three main ways of polarizing a radio or light wave are linear, circular and elliptical. Linear polarization can be vertical or horizontal.

Until 1864, when Clerk Maxwell showed by his equations that optics and electromagnetism are related, no one appreciated this. The two were thought to be quite separate but we now know they are linked as shown by Table 1.2, Chapter 1.

Since light is an electromagnetic wave, a demonstration of polarization can be made with two pairs of sunglasses. (The sun, as a source of electromagnetic waves, is like a broadcasting station.) If

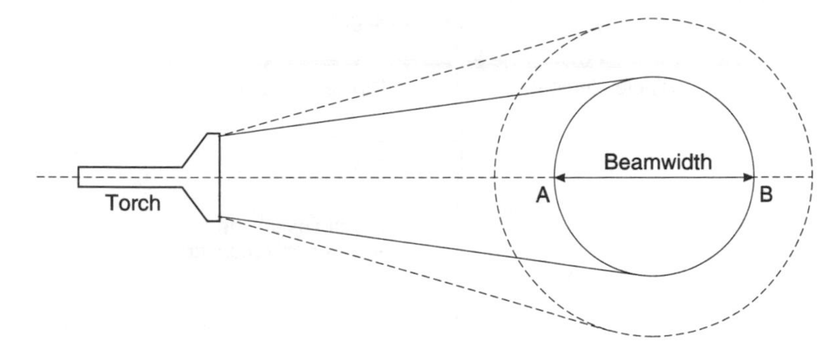

Fig. 5.1 Pocket torch, demonstrating beamwidth of an antenna

two of the lenses of the sunglasses are held opposite each other, light will pass through them unless the planes of polarization are perpendicular to each other. If one lens is slowly rotated relative to the other, light will begin to pass through again. Note: the glasses must be polarizing.

With radio waves, cross polarization occurs when the planes of polarization are not aligned with each other. Put simply, for dipole or beam antennas, a vertical antenna receives or transmits vertical waves, while a horizontal antenna receives and transmits horizontally polarized waves. Any receiving antenna receives the utmost from a wave polarized in the same manner as the antenna.

The two antennas mentioned below (dipole and beam) can be used in direction-finding and so they could help in locating a source of interference. If an antenna is directional it can of course help in avoiding a source of interference, merely by being aimed away from the source. Any antenna can be used in direction-finding provided that it is short enough to rotate easily.

Dipole

One of the simplest and best known antennas is the dipole (Fig. 5.2), a straight horizontal metal rod of which the two halves are connected at the middle of its length to the receiver through a coaxial cable. Its overall length may be a quarter-wavelength which means that for hand-held direction-finding the frequency should not be below about 70 MHz. This is because at 70 MHz the wavelength is about 4 m, which could allow a dipole to be only 1 m long. For higher frequencies the dipole can of course be shorter. But at lower frequencies it would have to be longer and could become impossibly cumbersome.

Transmitting antennas have to be suited to the wavelength they

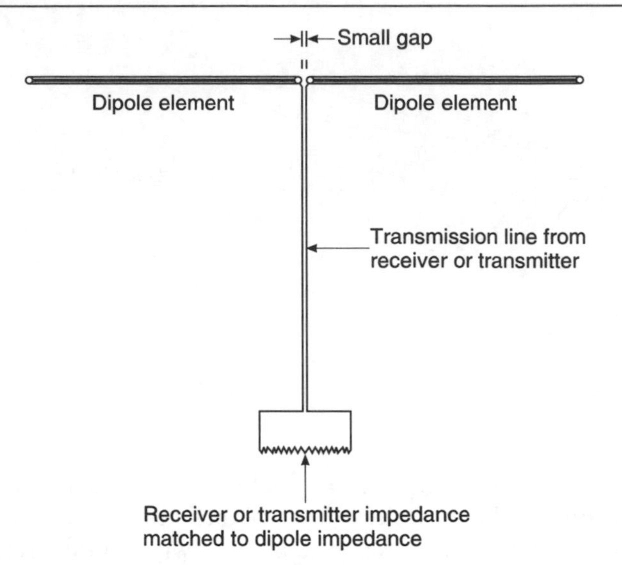

Fig. 5.2 Simple dipole antenna

transmit. They must be of resonant length, which, for a dipole, means that it should be a half-wavelength long. At high frequencies, with wavelengths of 30–300 metres, this is hard to achieve but it is easier at higher frequencies. At lower frequencies the enormous lengths are not practicable and transmitting antennas are then vertical.

For a dipole receiver the ideal wavelength is twice the overall length of the dipole but for convenience the dipole may be a quarter-wavelength long.

Beam

A more complex antenna, the beam, is much heavier and shaped like the capital letter H, sometimes with only two verticals but often with as many as six. The horizontal of the H is regarded as the antenna axis. For these two antennas the direction of best reception is, as one would expect, normally perpendicular to the axis of the beam, or to the length of the dipole.

Beams, like most antennas, are specialized for use at particular frequencies, so before use one must be sure that the beam in question is suitable for the frequency of interest. Beams are used to improve the directivity of the radiation and create a narrower beamwidth. Each additional rod in the beam tends to focus the radio wave, just as one uses mirrors to focus a beam of light in a particular direction.

Almost any antenna works best if it is well above ground, whether on a roof, a hill, a mast or slung from a kite or a balloon. The height minimizes the shielding effects of hills, buildings or other obstacles. Kites or balloons should not be flown over buildings, roads or, most seriously, overhead power lines. It is also important, if a thunderstorm threatens, to bring down the balloon or kite so as to disconnect the antenna. Lightning, like any electrical discharge, is attracted by sharp points and all antennas have sharp points, so disconnect antennas when lightning threatens!!

Those who use kites in a thunderstorm should remember that although Benjamin Franklin was lucky and survived, his Russian contemporary Richmann was killed while copying Franklin's experiment.

Testing at the pre-production stage

Probes can help to avoid manufacturing expense and delay if they are used for trying out printed-circuit boards at a pre-production stage when design changes cost little and are easily made. The probe can give a clear indication of leakage through a shield or of unwanted, excessive emissions from a circuit. Naturally one should be clear whether the probe needed is for magnetic or electric fields.

The small loop antenna (Fig. 2.6, Chapter 2) of 30 mm (1.2 in.) diameter can sense magnetic fields. The loop is of semi-rigid coaxial cable. The outer tubular conductor of the coax is sawn through at one point as shown. At one end of the antenna ring a 50 Ω resistor is connected to the central conductor of the coaxial cable. The other arm of the coax ring leads to the 50 Ω measuring system. Figure 5.3 below shows a probe to detect electric fields, designed to work as a capacitance.

Wavelength and frequency

To understand antennas it is essential to repeat a small amount of theory. Table 1.2, Chapter 1, shows that electromagnetic waves include radio, visible light, ultraviolet and infrared (heat) waves. Wavelengths are calculated from the universal formula that applies to all waves, whether they travel in water, air or any other medium:

$$\text{Speed (m/s)} = \text{wavelength (m)} \times \text{frequency (Hz)}.$$

The speed of all EM waves is about 300 million metres/second (300 000 kilometres/second). Frequencies, when measured in megahertz (million cycles/second, MHz). thus give the wavelength by the following simple equation:

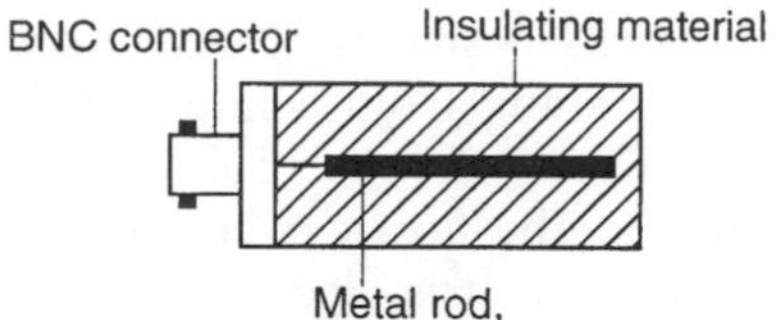

Fig. 5.3 Home-made probe designed to work as a capacitor, and able to detect an electrical field (after Goedbloed, 1992)

wavelength in metres = 300 divided by frequency in megahertz

At the radio frequency of 30 MHz (30 million cycles per second) the wavelength is thus 300/30 = 10 m. At twice this frequency it is 5 m and at half this frequency it is 20 m.

Table 1.2, p. 6 lists wavelengths at different frequencies and their common names (VLF, LF, HF, VHF etc.), but quarter-wavelengths are almost as useful to know because a piece of metal a quarter-wavelength long can act as an antenna both to receive and transmit this wavelength, possibly causing interference.

The wavelength at 30 MHz frequency is 10 m. At or below this frequency, measurements to test the EMC compliance of a device are usually made in the near field, at less than one-sixth of the wavelength from the source of the broadcast or radiation, as in the far field the measuring distance could be excessive. For frequencies of 30 MHz or less the authorities require measurements of the magnetic field, which have the advantage also of being fairly reproducible.

Suitable antennas for measuring a magnetic field are shown in Figs 2.5 and 2.6 of Chapter 2, Regulations and enforcement. The loop diameter of Fig. 2.5 is 0.6 m (2 ft) and this type is prescribed by many authorities for testing. A much larger spherical arrangement based on three such loops of coaxial cable intersecting perpendicularly and of 2 m diameter or more is beginning to be used for tests. For this antenna the equipment under test can conveniently be placed inside. There is also a trend to increase the diameter from 2 to 4 m.

A broadband antenna is one that receives or transmits a wide range of frequencies. Different broadband antennas cover the ranges: 25–300 MHz, 150–1000 MHz, and for microwaves 1–18 GHz.

Any piece of wire or metal if correctly oriented, even a heating or cooling pipe, can act as an antenna. The human body can also be an antenna as radio enthusiasts know. The body contains a high percentage of impure water as well as many tubes containing such electrically-conducting liquids.

Near field and far field

An EM wave has two parts (components), a magnetic and an electric wave. They are perpendicular both to each other and to their direction of travel and move away from their source at 300 million metres per second as shown by Fig. 1.3, Chapter 1. Far from their source, in the 'far field' they become a 'plane wave' and they behave differently in the near field and the far field. For writers in English the boundary between the two fields is at a distance from the source equal to the wavelength divided by 2π, usually taken to be 6. In other languages writers do not necessarily agree with English writers about the boundary between the near and the far fields. The German writer Ewald Kalteiss (1996) uses 10 times the wavelength instead of one-sixth. At such frequencies as power frequency, wavelengths are 6000 km in Europe and 5000 km in America. For such long waves the difference between English and other writers is unimportant.

Near field

In the near field, decisions about shielding, discussed in Chapter 9, are based on whether the field is mainly electric or mainly magnetic. Figure 5.4 depicts some of the terminology. Electric sources include spark gaps and other causes of arcing, like switchgear. A mainly magnetic field often comes from a heavy current flow through an inductor, e.g. a transformer.

Fields

A field is a region in which an object is affected by a similar object some distance away without any visible link between them. Science recognizes three types of field that can exert a force through space: electric, magnetic and gravitational fields. We are all too familiar with the gravitational force acting on a ball or other object which drops when its support is removed. Similarly a charged particle (e.g. electron) is affected by another charged particle nearby because of the region around it which is its electric field. If a charged particle within the field of another charged particle has the same polarity (positive or negative) then the particles move away from each other. If, on the other hand, they carry opposite charges they will be attracted to each other.

A magnet or compass needle is a magnetically polarized body. It has an associated magnetic field which affects and is influenced by other

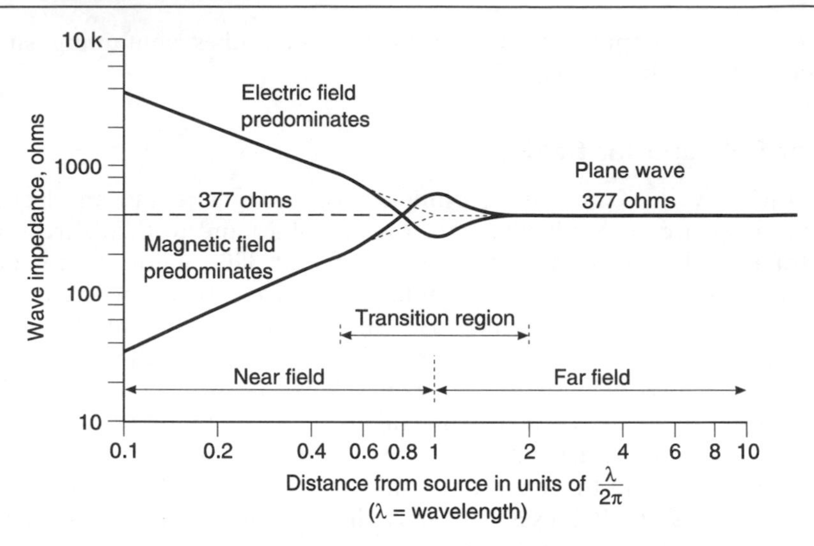

Fig. 5.4 Wave impedance in the near or far field (after Williams, 1992)

magnetically polarized bodies such as the Earth. The Earth's magnetic field causes the compass needle to align to magnetic north.

From the lay person's angle, the main distinction between near and far field is their variation in field strength with distance from their source. Ordinarily the strength of the far field varies inversely with distance from its source. Thus a field that is 2 km from its source has a strength equal to a half of that at 1 km from the source. In the near field, field strengths vary much more intensely. According to some writers they vary in inverse proportion to the cube of the distance; in other words, a field at 1 km could be eight times as strong as one at 2 km.

The wave impedance, indicated in Fig. 5.4, is a convenient figure for the designer of a shield because it shows him or her how to design the shield for an electromagnetic wave with the known value. It is the ratio of the field strengths, electric to magnetic: thus $V/m \div A/m = V/A$ (ohms).

In the near field the wave impedance depends mainly on the type of source, whether it is magnetic or electric. A source with high voltage and low current will generate a mainly electric field of relatively high impedance. If the source has a high current it will generate a strong magnetic field. If its voltage is low the field will be mainly magnetic and the impedance in the near field will be low, as can be seen on the diagram.

Chapter 6
Electrostatic discharge: sparking

Helpful electrostatic devices using mains electricity are found in many modern processes, including photocopying and the precipitators that remove dust from the smoke of power stations. This chapter, however, deals not with them but with the harmful effects of unintentional currents at high voltage from an electrostatically charged object such as the human body.

Electrostatic discharge (ESD) which is harmful is generated by electric charges on surfaces, either of conductors or of insulators, including chairs, tables and people. If the surface of an insulator becomes charged, the charge stays. If the surface of a conductor is charged, the charge can be removed or at least reduced by touching it with another conductor, but an insulator cannot be discharged in this way. Charged insulators are thus one of the main causes of ESD and its damage.

ESD is rarely obvious. But it always has a sequence, usually as follows:

1. An insulator is charged, either by contact or rubbing or induction (nearness to another charged surface).
2. The insulator charges a conductor, which may be the human body, by induction.
3. The charged conductor, approaching another conductor, discharges itself, sometimes by arcing.

Arcing cannot usually take place unless the voltage related to the air gap is more than 30 000 V/cm of gap. In the jargon of electrical engineers the breakdown electric field strength of air is 30 000 V/cm. People do not feel or see an ESD from the body unless the finger discharging the current is at more than 3500 V. Normally people cannot be raised to more than 25 000 V but they often reach half this level.

Conditions favourable to ESD are common in information technology. The computer and its peripherals heat the air around them and reduce its relative humidity appreciably, often below 50%.

A complication with ESD is that equipment which functions well in a wet climate like the UK may not work at all in a dry climate such as north Africa, or even in the dry air of a frosty winter in Europe. Damp air encourages charges to flow away while dry air is more insulating.

Human capacitors

In ESD the human body acts as a capacitor (100–200 picofarads (pF)) which, until release, stores the electrical charge at a high voltage. But a change of posture alters the capacitance as the body approaches or recedes from surrounding objects. Rising from a chair, for instance, reduces the capacitance but raises the voltage of the body. Any charge concentrates on extended parts of the body such as an outstretched hand.

ESD damage most easily happens when a device is directly touched, during either manufacture or servicing. Humans usually feel no electric shock from the common discharges they cause at potentials below 2000 or 3000 V. Even at 2000 V they feel only a prickling. Equipment should be designed so that vulnerable metal parts are not touched by people. For example, the metal shafts of potentiometers and the contact pins of computer connecting plugs can sometimes be at least partly insulated from human touch.

The damage done by a charged human body comes from its powerful electrical discharge of several amps at some thousands of volts – enough to melt metal or silicon. But serious electromagnetic interference also arises from the momentary high frequency sparking. The sparking current takes 1 μs or less, but fluctuates at several MHz. The pulse rises to a maximum in 1 ns (nanosecond), and the whole pulse usually occupies less than 100 ns (0.1 μs, a tenth of a microsecond).

Causes of electrical charge

The two main ways in which electrical charges are produced are either by triboelectricity or by induction (closeness to another charged surface). Triboelectricity is the charging of surfaces of dissimilar materials by rubbing together, followed by withdrawal. Plastics, including nylon, packaging film, coffee beakers and adhesive tape, are all sources of ESD. The human body is a conductor surrounded with a thin insulating skin and charging of the human body is usually a combination of triboelectric and induction charging.

Triboelectricity accumulates a charge when someone walks in rubber-soled shoes over a nylon carpet. In dry air the shoe soles and the carpet acquire charges of opposite sign because of the friction. Facing the soles of the feet, the skin, also an insulator, is charged by induction from the shoe soles. The feet are connected to the rest of the body by an electrical conductor, the blood, which spreads the charge throughout the body, concentrating it in pointed areas like the fingers. Induction charging results when an uncharged conductor is brought near a charged body, that is, into its electric field.

A third type of charging is by conduction. If two ungrounded conductors, one of which is electrically charged, come into contact, the charge on the first has a new potential and the second conductor acquires a similar charge. The charge is shared between them.

In the absence of contact, an ESD happens when the potential difference between two conductors exceeds the breakdown strength of

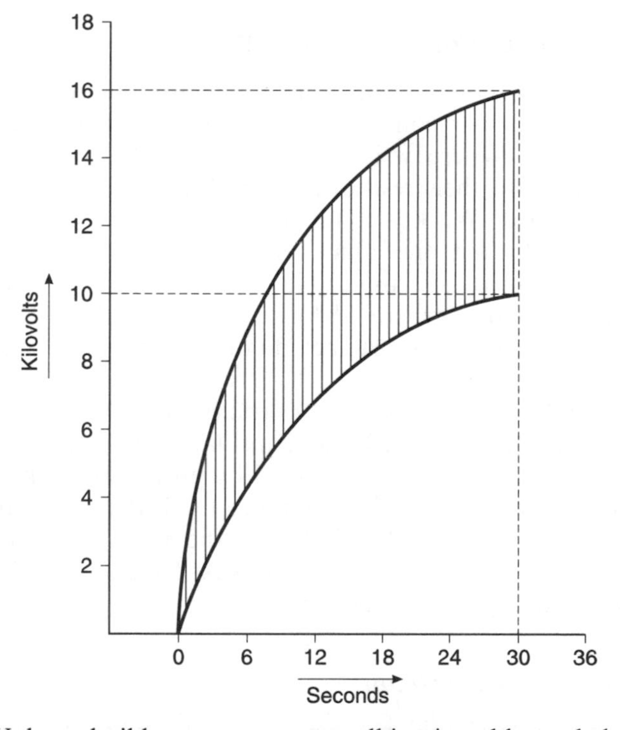

Fig. 6.1 Voltage buildup on a person walking in rubber-soled shoes on a polyamide carpet. After some 30 s, reaching 10 000–16 000 V, a balance is achieved between the rates of charge and discharge (after Marshman, 1995)

air. It occurs usually at a sharp point like a fingertip because this is where the electric field is strongest.

Three types of ESD failure

Of the three common sorts of failure caused by ESD, the least severe is a soft error. It can be corrected by re-setting the appliance (pressing some computer keys) and usually involves only a change of state of one or more computer bits. Hard failure is more serious, the failure cannot be corrected and a part must be replaced. The third type is latent failure, which occurs only after a delay, possibly from a radiation effect of an ESD.

People have to take care to avoid touching semiconductors because ESD damage to such a device from a person who has become electrically charged can be extremely serious. People are often unwittingly charged up to 10 000–15 000 V. Testing for an appliance's ability to withstand ESD is laid down by IEC 801 Part 2, 1991.

Laboratory precautions against ESD

People working in contact with semiconductor devices often wear a conducting wrist strap connected to ground, which keeps them at ground potential. Where wrist straps are relied on they should be regularly tested, and their resistance to ground should be about 1 MΩ (one million ohms). Weakly conducting clothes or flooring materials also help by allowing the quick release of any charge acquired. During the landing of an aircraft, its conducting rubber tyres release the electrostatic charge generated by its contact with air. Similarly, in the laboratory, conducting rubber floor covering or mats can be useful.

Materials that generate static electricity must be kept out of a workshop concerned with ESD, and insulators in particular must be avoided since they cannot be discharged by grounding. Ionizing the air makes it more conductive and encourages the leakage of any electrical charge. Maintaining a high humidity has the same effect. The relative humidity of the air should be kept to at least 50% and preferably 75% but ionizing is thought to be better.

Antistatic packaging: bench tops

Packaging methods should not generate static when the appliance is removed from its packaging. Bags that are made of pink plastic with some electrical conductivity are used for packing semiconductor units. The most effective antistatic bags have a built-in continuous metal foil. Antistatic materials attract moisture and allow charges to leak away.

Bench tops for people working on ESD are best made of any material described as static-dissipative, that is with less conductivity than a metal. Such materials ensure a slow and less dangerous route for ESD than a conductive metal bench top.

The damage possible from ESD provides an incentive for manufacturers to 'harden' their products against it, namely, to make them immune to ESD damage. A protection network may be built into the product, including an easy path to ground (earth) for the ESD. Gas tube surge arrestors, if specially designed for the two, can protect both against lightning and ESD. Spark collectors also help with the selective routing of ESD to the ground. The spark collector should however be carefully placed, well isolated from sensitive circuits and connected to ground through a known small impedance so as to keep radiated EMI under control.

Probing for ESD weak spots

A kitchen gas lighter, that works by the piezoelectric effect, can be used to test for the areas of digital or similar systems that are most susceptible to ESD. Each newly-designed printed-circuit board can be checked as soon as the first is built and before the production stage is reached. Corrections can then still be made at relatively low cost.

Because of the different possible types of electrical discharge, especially pulses in ESD, it has been found that an appliance can pass a test at 10 000 or 12 000 V but fail an apparently similar test at only 3000 V.

To check the vulnerability of an appliance against ESD, the IEC in its Standard 801-2 of 1984 describes its 'ESD gun'. This ESD generator includes a device for generating high voltage, as well as resistances, capacitances and a discharge electrode. It is designed for peak discharges of 9 A at 2000 V, rising to 70 A at 15 000 V, with a tolerance of $\pm 30\%$. These very high values may seem less frightening when one realizes that they rise to their maxima in 5 ns (nanoseconds), fall to half the values stated in 30 ns and to zero in about 60 ns.

Under IEC 801-2 the EUT is placed over but insulated from a ground plane of 0.25 mm thick copper or aluminium sheet that is large enough to project at least 10 cm beyond the EUT on all sides. The insulation over the ground plane has to be 10 cm thick. The ground plane is connected to a cable that must be 2 m long. Any spare cable must be carefully bundled so as to reduce inductive loops to a minimum.

Other sorts of sparking

Arcing and glow discharge

A quite different subject from ESD involves arcing and glow discharge. Strictly, arcing is current flow across a gap, accompanied by metal vapour discharge from contacts under an increasing electrical field, although people usually think it is the same as sparking. To prevent arcing, the rate of rise of voltage at contacts must be kept below 1 V/μs. However, it can occur at a voltage much below that needed for the glow discharges mentioned below. Electrons leave the highest or the sharpest point on the cathode, heating it often to a few thousand degrees C, enough to vaporize the metal. This electron flow is called field emission but as soon as the metal has melted, the arc consists of metal vapour. For noble metals such as gold, silver or palladium, the minimum arcing voltage varies from 9–18 V and the minimum arcing current from 0.4–0.8 A. Because of the characteristics of AC, contacts rated at only 30 V DC can tolerate 112 V AC. AC goes through zero voltage at twice its supply frequency, and the arc can be extinguished every time this happens.

Glow discharge, also called corona or Townsend discharge, is a current flow between two contacts separated by a gap when the gas between them is ionized. With air at standard temperature and pressure and a gap length of 0.075 mm, glow discharge takes place at 320 V. With a different gap, either shorter or longer, more volts are needed but, after discharge begins, the necessary voltage falls. To avoid glow discharge, the voltage across the gap should be kept below 300 V. As already explained, arcing can occur at lower voltages.

Sparking at switch contacts

A subject related to arcing and glow discharge is sparking at switch contacts. It generates serious EMI but also damages the contacts so that they eventually need replacing which is expensive, apart from any stoppage of profitable work while they are changed.

Contacts are ordinarily described by their ‘rating’, that is the resistance which they can tolerate at the stated maximum voltage and current. Even at that rating there is arcing both on ‘make’ and on ‘break’. Some contacts with an arduous duty cycle have a special, high rating for an inductive load.

Severe ratings exist for electric lamps and motors, both with a high starting current. For incandescent lamps the starting current is 10–15 times the rating and for electric motors 5–10 times the rating. Capacitors also draw high starting currents and they have to be restricted by the series resistance of their circuit.

Chatter or bounce

'Chatter' or 'bounce' of contacts which will not close immediately is a serious problem, resulting in repeated arcing, and radiation of high-frequency interference. Ten or more makes and breaks may occur each time with sparking, loss of metal at the contacts and annoying interference for neighbouring circuits.

Any sudden interruption of current through an inductor is a source of damaging spikes of current together with serious interference. The field coils of electric motors are naturally inductors and when the motor is switched off the energy stored in the field coils is released, causing sparking at the switch. Without protection the switch will not last long. Even a small 25-volt DC power supply can generate a spike of 500–5000 V when an inductor is suddenly switched off. Theoretically the voltage thus produced is infinite but the level is reduced by the sparking.

Protection of circuits against sparking

With such inductive loads, circuits can be protected against these heavy over-voltages by connection of several possible subcircuits, which usually involve insertion of a resistor, a capacitor and sometimes diodes.

The protecting network can be connected either across the inductor or across the vulnerable contacts and a varistor may be used. Varistors have a resistance that diminishes when the voltage across them rises so that the power loss is thus less than with a normal resistor. Chapter 10, Transients and protection against them, has more on varistors. Many other electronic protection methods exist, including the connection of a capacitor across the contacts, although this can be problematic, especially if the contacts bounce.

If an inductive load is controlled by a transistor switch the transient voltage generated by the inductor should not be allowed to exceed the breakdown voltage of the transistor. One way of limiting the voltage across the transistor is to connect a diode across the inductor. The diode should be as close as possible to the inductor to restrict the loop area of the wiring and thus the unwanted interference that could result from a large loop.

To restrict the damage done by the starting current of a small electric motor, one or more ferrite rings may be placed on the power conductors to the contacts. Ferrite rings limit the high frequencies of EMI but do not affect the steady-state current.

Organizations concerned with protection from ESD

ESD testing or hardening interests many bodies including the following:

BSI	British Standards Institution
DIN	German standards body in Berlin
IEC	International Electrotechnical Commission
ECMA	European Computer Manufacturers' Association
ANSI	American National Standards Institute
IEE	UK Institution of Electrical Engineers
IEEE	US Institute of Electrical and Electronic Engineers
EOS/ESD A	Electrical Overstress/ESD Association

Chapter 7
Susceptibility

First, an explanation of 'susceptibility' (the opposite of immunity). A victim device is susceptible if it experiences interference. But devices can be and often are both guilty culprit and susceptible victim at the same time.

Although it may interest EMC specialists less, the general reader may like to know that humans also can absorb energy at radio frequency. They absorb the most when the length of the body is equal to 0.4 times the wavelength of the field, and the body is parallel to the electric field. Thus, if the field is vertically polarized, the body has to be erect to receive the maximum radiation. Seventy megahertz is the most favourable frequency for the human body to receive RF radiation, because: speed of light (300 000 000 m/s) divided by the frequency (70 000 000 cycles per second) equals wavelength (about 4 m) and 0.4×4 m $= 1.6$ m, the average height of a human.

Reverting to EMC, treatment to prevent interference should, where possible, begin at the source of the interference. In support of this priority, if the intermediate stages of transfer are tried first, the interference may increase, but if the source can be reduced there is no need to worry about how the disturbance was transferred. Thus the source of the interference, where known, must be treated first.

Electrostatic discharge, discussed in Chapter 6, is a concern both of EMC and of susceptibility because it is not always easy to make an appliance immune to ESD. Most electronic units are susceptible to it.

National telephone systems

Telephone systems are sources of interference and sometimes also suffer it, but they are extremely varied and only a general introduction to them can be given. Each country's telephone system carries speech, telegrams, data and programmes for broadcasting on radio or TV. They use combinations of copper wire, optical fibre, satellites and microwave relays using line-of-sight connections. Private telephone systems owned by gas or electricity companies or railways use similar methods. These organizations, like the police and the fire and ambulance services, also use land mobile radio services (LMR or PMR) which may sometimes be connected to the country-wide telephones. They existed long before the modern mobile phones using cellular radio.

Microwave transmission needs relay stations at intervals decided by the line of sight. The relays operate automatically so that each message received is amplified before retransmission. The most modern relays use digital techniques which have the advantage that any number of relays can be used successively without failure. Analogue techniques at each relay unfortunately accumulate the faults from all previous relays so that the length of transmission is limited by the unavoidable progressive deterioration of the message.

The mechanical relay-operated devices that ran the telephone exchanges until the 1980s were superseded by much smaller assemblies of semiconductors with their computers and software. The speed of the semiconductor devices, their flexibility and small size were far too attractive to be ignored at a time of fast-expanding demand, so semiconductors came in and electromechanical systems went out.

It must be remembered that the old systems tolerated much wider ranges of temperature, humidity and EM interference than is possible for semiconductors using tiny currents and small voltages of 5 V DC compared with the 48 V of the old systems. Unfortunately the old systems often generated interference, but they did not collapse disastrously, which is possible with computers when overloaded.

Control of telephone exchanges by computer therefore has to be duplicated with at least one spare backup computer provided with automatic switchover in the event of failure or overload. The essential spare set or sets, although expensive, can help by taking over demand that is increasing too fast.

Susceptibility limits

When the strength of the electric field where an appliance is to be installed is much less than 1 V/m it is unlikely to suffer severe interference. For some purposes, however, a higher level, 3 V/m, has been taken as the electric field below which appliances may be immune from interference.

The strength of an electric field depends on the power of the source transmitter, and its distance away. A person must be at least 2.6 km away from a powerful 2000-kW radio or radar transmitter before he or she experiences an electric field below 3 V/m. For a much less powerful 50-kW TV broadcasting station the distance needed is some 400 metres. For an even smaller 200-W amateur broadcasting set the distance is 26 metres and for a 50-W land mobile radio it is 13 m before the electric field drops to 3 V/m.

In the summer of 1995, a new type of interference was reported by hospital authorities who complained of the digital mobile phones available in the UK since 1993 and known as GSM. Several types of hospital personal appliance were affected by GSM mobile phones, including heart pacemakers, lung ventilators, hearing aids and wheel chairs. Most of these mobile phones have a maximum output of only 0.8 W although the most powerful reach 8 W.

Where signals are displayed on an oscilloscope screen and the interference is seen alongside the desired signal, both can be recorded for study and where necessary used as evidence to prove the guilt of the emitter.

The two main susceptibilities to interference in EMC come through cables by conduction or through space by radiation. Radiation investigations can involve a complication, the use of an anechoic room. (More on anechoic rooms in Chapter 2.) Whenever an appliance is connected to cables outside it, that are longer than the largest dimension of the appliance case, these cables become the greatest contributors to both radiated and conducted susceptibility. The reason for this is that cables act as antennas to pick up any interference being radiated.

Common-mode and differential-mode currents

From the viewpoint of EMI, cables and any other conductors can carry two types of current. One type, the unwanted common-mode (CM) current, can produce serious interference. The other type, differential-mode (DM) current, carrying a wanted signal, generally does not because it flows along paired conductors that are close together. One is an outward conductor, the other is its return. Although it is possible for interfering currents to flow in these conductors, they are usually less disturbing than CM currents because the two DM wires have currents flowing in opposite directions, and their fields largely cancel each other out.

The unwanted, stray, CM current on the other hand flows in the same direction along all conductors in the cable, including the cable shield. It returns through the ground connection. The ground connections may have large loop areas which produce serious interference even with a very small current. Although usually smaller than differential-mode currents they create much more serious interference because they are not found in paired wires that cancel each other's radiated fields. DM conductors unless they are far apart and not twisted together, contribute less interference than CM currents because CM currents have large loops to ground.

In the USA, the FCC's susceptibility requirements for radiation are much stricter than those of the CISPR for Europe. For dwellings, the FCC requires compliance tests in a frequency range from 30 MHz to approaching 40 GHz. The CISPR also starts at 30 MHz but its upper limit is only 1 GHz. In the USA the distance for measuring emissions is 3 m for the dwellings of class B and 10 m for class A (industrial and commercial) products. CISPR 22 requires 10 m for class B and 30 m for class A products, which is of course much more lenient.

For these frequencies there are the following wavelengths: at 30 MHz the wavelength is 10 m; at 1 GHz the wavelength is 30 cm; and at 40 GHz the wavelength is less than 1 cm. With these wavelengths and the distances listed, it is clear that both the near field and the far field are involved. Emissions in the near field are much more complicated than those in the far field. In the far field a fair assumption is that the magnetic and electric fields vary inversely with the distance from the source of the wave. Very different assumptions exist for the near field, where the reduction in strength is proportional to the cube of the distance from the source.

Common-mode currents cause the most interference even although they are smaller than the differential-mode currents. Without the complication and delay of hiring a screened room it is possible to measure common-mode current with a home-made current probe which is fairly easily calibrated by taking readings with it while currents of known value are passing through the conductor under investigation.

Figure 7.1 shows a current probe. It is built of a halved, hinged ferrite toroid which can easily be clipped over any cable. Four turns of wire round the ferrite lead to the voltage measuring instrument chosen by the experimenter. This type of probe is inexpensive, easy to use and gives repeatable results for RF currents from microamps to 20 A and occasionally more. But at very high currents the ferrite can become saturated, resulting in false readings. The normal operation of the

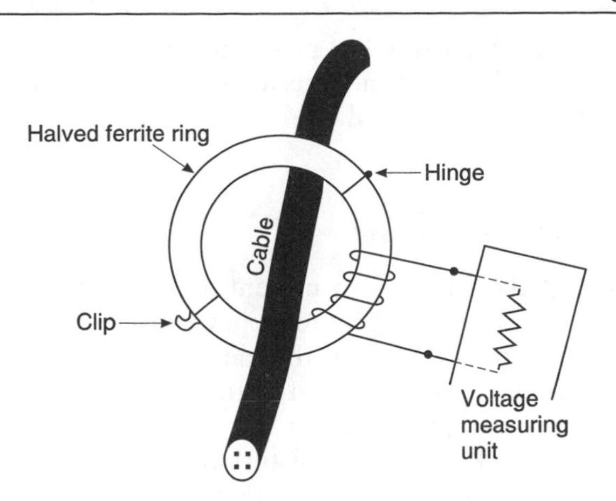

Fig. 7.1 Home-made current probe

probe is to function as a transformer, using the conductor under test as its primary coil. A current probe can be specially made to inject RF current into a conductor and such probes may be made in sets to suit particular frequencies. For EMC testing, however, the Line Impedance Stability Network (LISN) mentioned below is more commonly used than the probe.

Susceptibility to conducted interference from power cables

Perfectly legitimate electrical vibrations within the circuits of an appliance may be transferred out of them along its power cable to cause interference to other power consumers elsewhere. The control of these conducted disturbances is as important as controlling radiation, perhaps even more so.

FCC limits on conducted disturbances extend from 450 kHz to 30 MHz while CISPR 22 has wider limits, between the frequencies of 150 kHz and 30 MHz. Both in Europe and in the USA, when an appliance is tested for its compliance with the law on EMC, a LISN has to be used. This black box provides a power supply allowing comparable test results from a theoretically constant but in fact ever-varying public electrical supply. All mains electrical supplies have varying voltages so the stability of the LISN helps in making comparisons. Normally one LISN is connected to each conductor of the power supply. The purpose of the conducted emission test is to measure the noise currents that come out of the appliance from its power cable. Although to the non-electrical person it might appear strange that the cable which supplies power to the appliance can also, in the opposite direction, inject interference to the supply, this unfortunate fact has to be accepted.

Line impedance stabilization network

The first objective of the LISN is to present a constant impedance to the appliance's power cable over the frequency range of the conducted emission test. The electronic noise on the mains power network varies from site to site, but the aim of the test is to measure only that noise which issues from the equipment under test (EUT). One aim of the LISN is therefore to block conducted disturbances that are not caused by the EUT. In other words the LISN should prevent the conducted noise of the mains power system from contaminating the measurements.

The LISN should present a constant impedance (50 Ohms) between the live conductor and the safety wire (yellow and green in the UK, green in the USA) and also between the neutral conductor and the safety wire.

The currents passing through the LISN may be separated into a wanted differential-mode component flowing out through the live conductor and returning on the neutral wire and an unwanted 'noisy' common-mode component flowing out through both live and neutral, as well as on any cable shields and returning on the safety wire.

Unlike radiated emissions, conducted common-mode currents can be as large as or larger than the differential-mode currents. Common-mode currents, if conducted, can be appreciable.

Where interference has to be prevented from entering an appliance or, on the other hand, disturbances in the appliance have to be prevented from polluting the mains, it may be worth considering the connection of a power-supply filter. Chapter 11 has more on filters.

Chapter 8

Impedance matching, balancing, tuning; noise and interference; car radios; direction-finding

All RF circuits need to be optimized, and there are different ways of doing this. Connected RF circuits should be matched or balanced to each other and the individual circuits also need to be tuned. Most people are aware that a radio set must be tuned so as to receive the frequency or wavelength as well as possible. Noise and other interferences with reception are mentioned later in this chapter.

Tuning

Tuning involves adjusting the capacitance or inductance of an AC circuit so that it resonates. A resonant circuit has the maximum response at the wanted frequency with the minimum of electronic noise. Resonance is achievable by tuning because capacitance and inductance behave oppositely with changing frequency. When the frequency rises the inductive reactance rises but the capacitive reactance falls. Tuning means increasing or reducing one or the other until the two are in balance, in other words, they cancel each other out at the required frequency.

Balancing

Balancing is a cost-effective noise-reduction technique which may eliminate the need for shielding or be used in addition to shielding. Its purpose is to make the noise picked up in parallel conductors equal, so that any noise is cancelled out. Ordinarily it means simply equalizing the impedances of the two wires. 'Balanced' usually implies electrically alike and symmetrical about a common reference point, usually the ground. Balanced circuits tend to emit less radiation and therefore less interference than unbalanced ones but no exact estimate can be made, especially above 30 MHz.

Impedance matching

To ensure that an RF circuit works most efficiently, with the maximum signal and minimum noise, the impedances of connected circuits should be the same, i.e. 'matched'. Thus the feedpoint impedance of an antenna should equal the impedance of the transmission line to it. Where there is a mismatch it can sometimes be corrected by connecting a balun between the antenna and the coaxial cable of the transmission line.

A balun (BAL UN) is a component that connects a BALanced to an UNbalanced impedance. Many different types of balun exist and they can be home-made. Another device is called an impedance-matching unit. At higher frequencies, when every resistor acquires stray reactances from capacitance and inductance, impedance matching becomes extremely difficult or impossible.

For FM radio or TV reception, disturbance created by wrongly wiring a receiver, is easily avoided by correct installation, especially of coaxial cable (Figs 8.1 and 8.2). The BBC recommends also that the antenna should be above any nearby antenna, in fact at least 1 m higher, without specifying how to eliminate any resulting disturbance at the lower antenna.

Sources of noise

Noise is electrical disturbances often of unknown origin, but usually less powerful and so less disturbing than interference. White noise is a name for broadband noise. It has uniform power over a wide range of frequencies. A common view is that noise is generated within the semiconductors of a circuit, but interference comes from outside it. The 'snow' on a TV screen is often noise from the semiconductors of the amplifier but it may also be caused by a bad coaxial connection, dampness, a 'dry' solder joint, or a mismatch.

Many sorts of natural noise exist, apart from lightning and the galactic noise of the Milky Way. 'Static' or 'atmospherics' are more common words for them. Man-made noise comes from many sources including photographers' electronic flash equipment, car ignition systems, switches, the brushes of electric motors or generators, digital

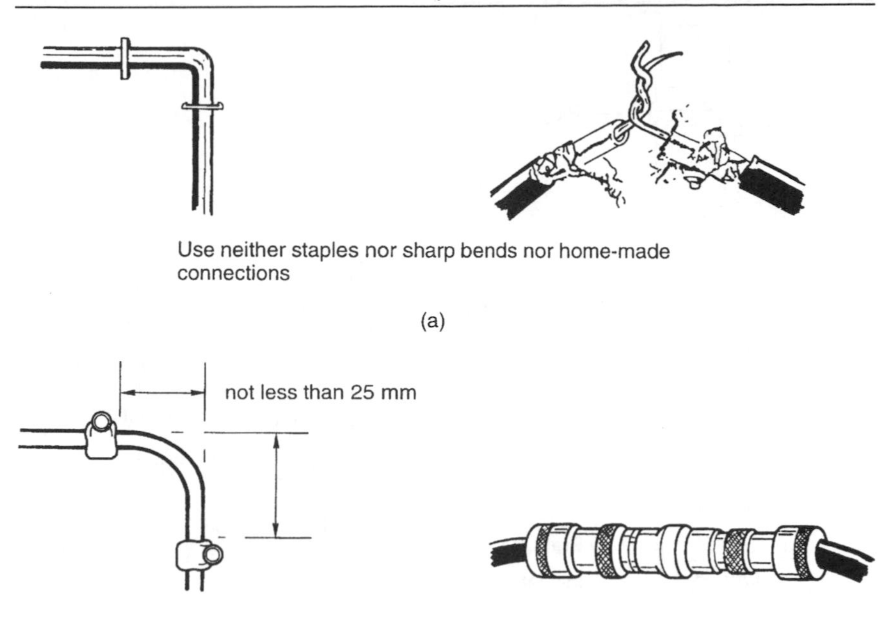

Fig. 8.1 How to fix a coaxial cable: (a) wrong; (b) right (source: BBC Engineering Information, London)

electronics such as computers, medical high-frequency devices and electronic igniters for cigarettes as well as radio transmitters, particularly CB radio or mobile phones.

In the specialized activity of signal recovery, where the signal may be submerged in noise, it can help to be able to distinguish the various types of noise.

Several types of noise originate in semiconductors. The first type has three names, thermal or Johnson or resistance noise. All resistors generate it but so do all components that have resistance, including diodes and transistors. However, pure inductors and capacitors do not, if pure, which none of them are. Flicker noise is sometimes called 1/f noise because its level is inversely proportional to frequency (f); in other words, at low frequencies it is more noticeable.

Although innumerable types of man-made noise exist, two special sources are worth mentioning, the first being due to the triboelectric effect, and the second due to electrolysis, sometimes called the rusty-bolt effect.

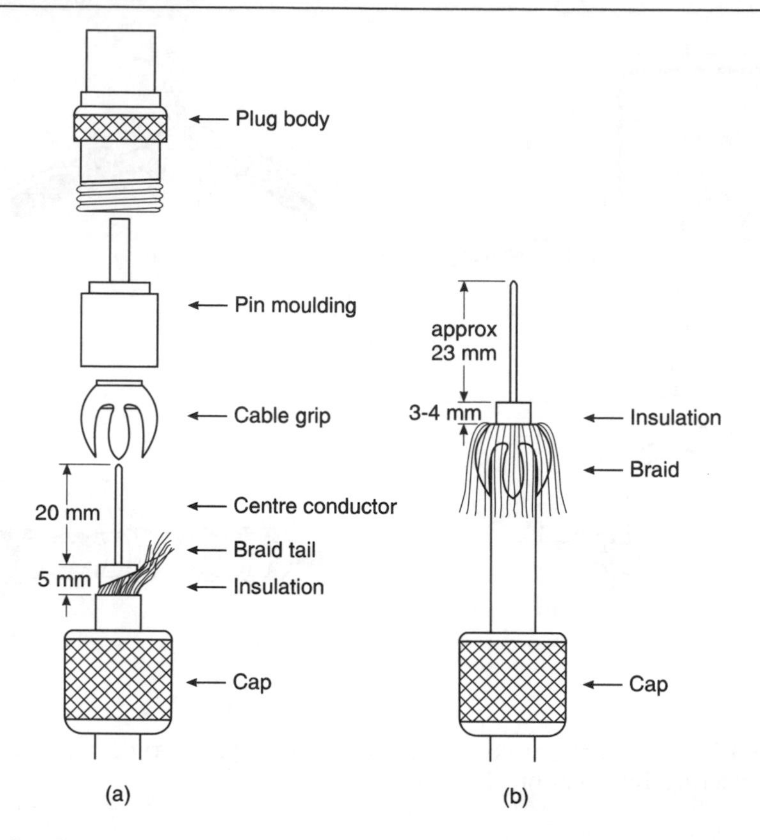

Fig. 8.2 Two ways of correctly fitting connectors for TV coaxial cable: (a) cut the braid tail to 5 mm max, wrap it round the 5 mm of inner insulator and push cable grip down on it. Do not allow braid whiskers to touch the centre conductor. (b) Tease out the braid, bend it back over the cable grip and cut it off (after Robin Page-Jones, *Radio Amateur's Guide to EMC*, Radio Society of Great Britain, 1992)

Triboelectric effect

In the triboelectric effect, two materials when rubbed together can acquire opposite electrostatic charges. The substance lower in the triboelectric series is negatively charged, while the substance higher in the series is positively charged. Thus any substance can be negatively charged if rubbed with one higher in the list, although positions in the series vary with the relative humidity of the air and whether the substances are dry or wet. See Chapter 6 for electrostatic discharge.

Triboelectric or electrostatic series (shortened and approximate)

Positive	*Negative*
rabbit fur	cellulose
glass	
mica	
nylon	
wool	
cat fur	
silk	
paper	
cotton	
wood	
amber	
resins	
metals	
sulphur	

The triboelectric series concerns cables, especially microphone cables since they often have to bend. If the insulator in a cable loses contact with its insulated wire, the insulator (dielectric) receives an electric charge, which creates a current and therefore noise in the circuit. Hence, if sharp bends in the cable are minimized, the disturbance is reduced. 'Low-noise' cables made for this purpose do not eliminate noise but they do reduce it.

This should not be confused with a different effect, caused by movement of a conductor in a magnetic field, mentioned earlier in the explanation of induction. This induces a voltage in the wire, resulting in current and consequently noise. Movement can be reduced by clamping the wire firmly, with cable ties and clamps.

Electrolysis

The rusty-bolt effect involves electrolysis. All metals have a place in the galvanic series, shown abbreviated in Chapter 12, Printed-circuit boards. Corrodible metals such as iron, steel or aluminium are high in the galvanic series and the noble metals, like gold, silver, platinum or palladium, are low in the series. Wetted contact between dissimilar metals (i.e. from different positions in the series) can corrode the metal that is higher in the galvanic series.

Impure water conducts electricity when acting as an electrolyte, allowing ions to pass from the anode metal which corrodes, high in the triboelectric series, to the cathode metal low in the series. The resulting electric current causes noise although the effect can be avoided if dissimilar metals are kept out of contact. This can be helped also by using plastics instead of metal. Grounding, another important source of noise, is dealt with in Chapter 12.

Figure 8.3 shows the relative strengths of noise at a receiver in a UK country district. As the diagram shows, atmospheric noise at VLF (very low frequency) is by far the biggest contributor. This is because VLF thunderstorm noise can travel long distances and the power of a lightning stroke rises to a maximum at VLF.

Check your TV set

Before considering the possibility of outside interference, the owner should first check that the TV set is in proper working order. When you notice noise on your portable radio, turn off the TV, and if the noise disappears the TV is the culprit and you must seek a professional to repair the TV. Spring-operated electrical units are likely to fail from old age or over-use, so first check all switches, plugs and sockets.

Check for intermittent power supply caused by faulty wiring. Then switch off the TV, pull the plug from the wall socket and, if you are qualified to do so, open the set to look at it in daylight, when other faults may show up. For example a discoloured resistor which has overheated may have changed its resistance. One's sense of smell is helpful also as a smell of burning varnish may come from an overheated coil or winding while a smell of urine may indicate corrosion of an insulator. Amateurs need a warning to be careful about the high voltages, even when disconnected, of up to 25 000 V in a

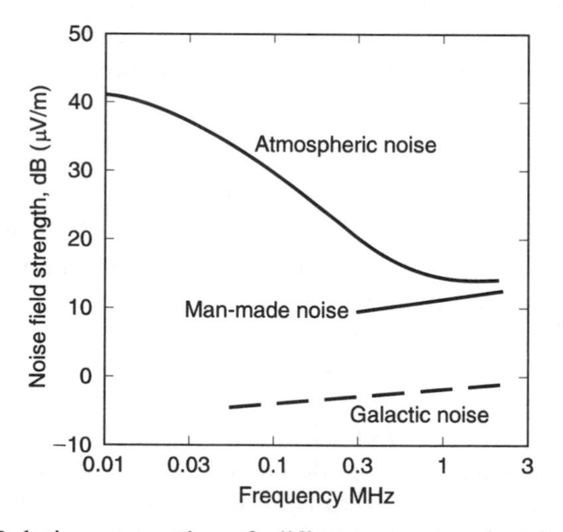

Fig. 8.3 Relative strengths of different types of noise in relation to frequency (after Hall and Barclay, 1989)

colour TV and a few thousand volts in a black and white TV, caused by charged capacitors in the set. For safety reasons do not open a TV set unless you are qualified to do so.

The components that are most likely to fail include resistors of high value, 100 000 ohms or more. Electrolytic capacitors can dry up in time and lose capacitance. White deposits around wire connections indicate corrosion, and probably faulty contacts. Finally, check any soldered connections, especially those that carry a high current or a heavy mechanical load. If all these checks reveal no obvious fault, only then should you consider looking for interference from outside.

A common-mode choke (Fig. 3.1, Chapter 3) reduces interference entering a pre-amplifier, amplifier, stereo, videocassette player, turntable, tuner or other unit through the power-supply cable. Keep some similar spare 2 m (6 ft) lengths of power-supply cable ready wound. If connecting one cable reduces noise a little, try adding a second wrapped ferrite ring. A professional adviser should tell you whether you are using the correct type of ferrite and the correct number of turns.

First check your own dwelling

A local authority which owned a broadcasting station circularized its neighbourhood with advice about locating electronic noise at home and found that as a result, complaints about radio or TV interference dropped by 30%. So it is worth while first to check one's own appliances. At your main fuse box, with your portable radio switched on and tuned only to the 'noise', switch off each circuit in turn by disconnecting its fuse. When you have found the guilty circuit the noise will stop. This ends the first part of your search. Reconnect the guilty fuse and go to the area which it controls. Once there switch off every appliance and light until you find the guilty one.

W. R. Nelson, in his *Interference handbook*, describes a fascinating search in which very troublesome interference had been traced to the kitchen. Everything in the kitchen was switched off but the disturbance continued and in desperation the investigator raised his portable radio to the central ceiling lamp which was of course switched off. The sound increased to a roar as the radio approached the ceiling and he eventually found a fault, in the electrician's connection to the ceiling light, which allowed leakage to ground (earth) from the lamp even when the power was switched off. An electrician was asked to correct the fault and this silenced it.

An inexpensive, portable, battery-operated, multi-band radio is the best tool for tracking interference whether on radio or TV. Although some people might think that a portable, battery-operated TV set would be better for searching out TV interference, this is not so and radio is better. The higher the frequency of the noise, the closer one is likely to be to the noise source. TV sets themselves are great producers of noise and so when searching for TV interference keep your portable radio well away from your TV set. To locate a noise source, adjust the portable radio to the highest frequency of disturbance and watch the TV picture. There is often a striking similarity between the bursts of sound on the radio and the distortion of the TV picture.

Types of interference on radio or TV

TV ghosting

A multiple picture can be caused by a reflection of RF waves from a tall object such as a building, providing a ghost alongside the legitimate picture. If the ghost moves it could come from a tree swaying in the wind or a moving crane. An inquiry to the local authority should reveal the existence of nearby building sites with tall cranes.

Uneven band of spots

Interference consisting of an uneven band of spots can come from the operation of a vacuum cleaner, a fan or an electric drill. If switching off the appliance makes the interference disappear, you have found the guilty source. All modern appliances are required by law not to emit interference but older ones may be guilty, and should not be used.

Short bursts

Short recurring bursts of TV interference may come from a fridge or freezer, timer, fluorescent light, thermostat or other device switching on or off. If an old thermostat or time switch is the cause, the expense of changing it for a new one that does not cause interference may be worthwhile. These disturbances may also be audible.

Herringbones

A herringbone pattern on the screen is likely to come from many sources, including a radio transmitter such as a CB radio, a radio taxi or other mobile radio and may also affect the sound.

Interference on a car radio

Before considering the installation of any sort of noise suppressor for a car radio, the owner should remember that a well-maintained ignition system produces less noise than a badly maintained one. Careful, painstaking, regular maintenance of spark plugs is the first step towards elimination of noise on the radio.

Sparks, not only at spark plugs, produce almost all electronic noise although the ignition generally produces most of the noise. Sparks come from the alternator slip rings and all devices with make-and-break contacts. Modern engines with electronic ignition are less noisy than older engines. Apart from ignition maintenance (spark plugs), there are several techniques of noise reduction which involve electronic circuits beyond the scope of this book.

Ignition noise is a popping sound that is synchronous with the engine speed, so is easily identified, as when the engine speeds up or slows down, so does the popping sound.

Alternator noise is a whine that begins only when the engine accelerates. If in doubt temporarily slip the alternator belt off or disconnect the alternator leads. If the noise continues the alternator is not the source.

Instrument noise from the fuel or other gauges is a hissing or crackling. To find out which one is guilty, disconnect each instrument separately until you find the source.

Wheel noise

While a car is moving, tyres and wheel noises from static electricity produce an irregular rushing sound in the loudspeaker. Naturally, this occurs only when the car is moving. To find out if you have wheel noise, switch off the engine and let the car coast downhill. Any electrical noise is probably from wheels or tyres. Wheel static, usually from the front wheels, comes from the insulating film of lubricant in the bearings. Formerly collector springs in the hub cap removed the static by grounding it but modern communication radios usually do not need such radical methods of noise elimination. The static discharge between car tyres and the road surface is worst on hot dry days and it can be eliminated, although not for very long, by wetting the tyres.

Testing for radio noise from the engine

The radio noise level must be tested with the engine off as well as with the engine running. Testing must take place well away from any external noise source such as other engines, high-voltage overhead power lines, neon lights, etc.

Switch off the squelch and any other noise limiter. Adjust the volume control of the receiver until the speaker gives a steady background noise and then select a weak station and reduce the volume until the station is barely heard. Turn off the engine and see if the noise level drops. If not, there is no interference from the engine but if the noise does drop your next task is to find out where it comes from.

It is unfortunately not always true that if you hear no noise there is no interference. Strong interference can lower the sensitivity of a receiver so that weak signals are not heard. Noise tests must therefore be based on a weak signal with only background noise.

The next point to decide is whether the noise is conducted from the power supply or other wires connected to the radio or through the antenna. Disconnect the antenna and then if the noise disappears it probably comes from the antenna and is radiated noise. If the signal disappears while the noise remains with the antenna disconnected, the noise must be conducted. First look for a loose connection on any wires to the radio and tighten them, then investigate further sources.

Spark-plug suppressors

Although a high voltage is needed for the spark to jump across the gap, very little current is needed. The quantity of interference is proportional to the current and so a high resistance in series with the plug therefore reduces the current and interference without greatly reducing the voltage. Suppressors may be effective but they may also impair the engine performance. If they do not work well enough, electronic filtering to reduce conducted noise and shielding to reduce radiated noise may be advisable as well as or instead of suppressors.

Filtering out engine noise to the radio

A feedthrough capacitor installed by an electrician can eliminate conducted noise from the battery circuit, but the battery cables also must be shielded (screened) to prevent them picking up noise from the alternator, ignition or any other guilty device.

Some noise eliminators that include a choke coil connected between the radio set and the battery are intended mainly for music and CD players rather than radio receivers.

Shielding (screening)

Chapter 9 deals more fully with shielding, but even with suppressors and filters there may still be too much engine noise on the radio. Do-it-yourself kits include shields to fit round plugs, as well as high-voltage shielded cables to the plugs and a shield round the electronic distribution unit. The whole may be sold as an assembled cable harness but even so, with full instructions may be laborious to install.

Bonding straps

The car's steel frame (chassis) is its ground (earth) to which every electrical unit must be effectively connected, since most electrical circuits in a car are completed through the frame. If the bonding is by welding, brazing, crimping or even soldering, the connections should be good, but further strap bonding by the DIY car owner helps to reduce radio noise.

A bonding strap usually has copper lugs at each end, joined either by broad flexible copper strip or by metal braid and the lugs are drilled for bolting. Metal parts that move relative to each other produce less interference when bonded together by straps. Otherwise, especially if they are dissimilar metals, they may not only corrode but also produce electrolytic current and consequently radio noise.

Some points which should be bonded securely to the steel frame include the engine corners, both ends of the exhaust pipe, the battery ground (earth) and the front and rear bumpers. There should also be good bonding between the distributor and the engine, between the alternator and the voltage regulator, and between the air cleaner and the engine block. A further discussion of bonding is given in Chapter 12, Printed-circuit boards.

Seeking the origin of domestic radio or TV interference

Radiocommunications agency

The Radiocommunications Agency (RA) of the Department of Trade and Industry, on request and on payment of £35 (in 1996), will diagnose interference on domestic TV or FM radio (but not AM) and only after the sufferer has already complained to his or her local supplier, whether dealer, service engineer or aerial contractor. As well as the cheque for £35 the application to the RA must include a statement by the local supplier that he or she has investigated the problem and cannot solve it. The relevant form must be sent to the RA with the signed statement from the local supplier, which confirms that the trouble is not caused by a faulty aerial. If the RA's investigator cannot diagnose the cause of the interference or finds that it comes from an illegal source, the RA will return the £35.

The RA's investigation does not cover interference on any appliance not designed to receive radio or TV, such as telephones, fax, answering machines, record players, or CD players. Car radios, cable TV and satellite TV are also excluded. However, interference on a telephone line should be reported to the telephone supplier who will correct the problem if it is not caused by faulty equipment.

The RA emphasizes that both fading and interference are common on long and medium wave (AM) especially after dark and that little can be done about these troubles, adding that FM has much better reception in terms of noise suppression and that a change to FM might help.

Private inquiry into radio or TV interference

Where someone wishes to pursue his or her own inquiries about radio or TV interference, this is quite possible without causing annoyance or even being noticed by anyone. Any antenna can be directional provided it is short enough to be easy to rotate, and can be used either in a handheld search or in a car.

Before beginning a search for interference it is best to try to inject some logic into one's inquiry by writing down some queries and answers to them.

- When does the interference occur and for how long?
- Have I connected a mains filter ? Did it solve the problem?
- Can any of the common sources of interference be ruled out? These include interference brought by the power supply or from a video-cassette recorder or leakage from cable television or a possible rusty-bolt effect, which could be from a rainwater gutter or a garden fence.
- Many devices that involve automatic switching, including touch-controlled lamps, have injected disturbing frequencies into the power supply which can affect radio or TV reception several miles away. Is mains interference conceivable?

If mains interference can be ruled out, a direction-finding search as described below may be feasible. For a particular radio, suffering from possible mains interference, the guilt of the mains can sometimes be settled by using a battery to operate the set but this is not always possible. If you wish to make sure that mains interference is excluded, it is helpful to connect a common-mode choke to the power supply of the set (Chapter 3, Power supplies, Fig. 3.1).

Because radio interference affects only the limited band of a sound broadcast, the search for it is easier than for a source of TV interference which may be anywhere in the vast area of some six megahertz – six million frequencies.

Direction-finding

For people determined to discover all the complexities of direction-finding (DF) there is a large American book, *Transmitter hunting, radio direction finding simplified*, by J. D. Moell and T. N. Curlee, with more than 300 well-illustrated pages. It was, in 1995, available from the Radio Society of Great Britain in Potters Bar but unfortunately the RSGB no longer stock this book although it might be available from radio libraries or by advertising for a secondhand copy in *Radio communication*, a monthly magazine.

A DF hunt can be highly enjoyable for the people who take part, but for those who do not feel justified in going to such lengths to discover the fascinating wizardries of the many Doppler DF devices with their multiple antennas, a simpler method is described below, for which the essential is an army type magnetic compass, apart from the usual portable radio.

After deciding that the mains are not producing interference, one can logically begin to consider outside sources. Confining ourselves to sound radio, there is one easily available search tool – a portable radio, for AM or medium-wave reception. Being battery-driven like all portables, it is highly unlikely to produce interference. It can be carried by the investigator on foot or in the car to find all those directions that at different points give maximum interference. During the search, the set must naturally be tuned to the frequency of the interference. If a large antenna can be carried, and it is small enough to be easily rotated, it should give better reception than the internal aerial of the portable.

The essential army type compass shows the magnetic north. When pointed towards the noise maximum it shows its magnetic bearing. For people unfamiliar with magnetic bearings a word of warning may help. No compass reading must ever be taken either near a car or near any other massive iron or steel object such as railway or tram rails. Iron or steel always falsify magnetic readings. Consequently before a compass reading is taken the compass must be removed well away from the car.

One may have to rely on the internal antenna within the portable. Ordinarily this is a ferrite rod with its length parallel to the length of the set. The best reception is therefore on a line perpendicular to the length of the set. If several maxima occur, the bearing of each must be recorded. But it may be comforting to point out that near the source of the interference the maxima occur most closely together. If a maximum is hard to find it may be easier to search for the worst reception, a minimum (null). Nulls, bearings at 90° from the direction of the noise, are found by rotating the antenna 90° from the supposed direction of maximum noise. The antenna should then point towards the noise.

All magnetic compass bearings must be treated with caution. The compass bearing may be wrong because of nearby iron or steel and the interference received may be only an echo reflected from a building, in which case its direction will be wrong. So an absolute minimum of two bearings is essential before the first preliminary fix is achieved. A fix is a point at which two bearings meet when they are drawn on a map of the district. It is the first attempt at a location of the interference. Once several maxima have been found by travelling along different routes, possibly yielding several different fixes, it should be possible to have a rough location for the guilty emitter, although deviations between the various fixes must be expected.

Faults in equipment

In searching for faults in their equipment, readers may find it helps them to build and use two inexpensive, do-it-yourself probes, one for magnetic fields, the other for electric fields. Figure 7.1 shows a probe for measuring electric current, and Figure 2.6 (p. 30) shows how to detect a magnetic field, even if small.

Chapter 9
Shields (screens)

A shield or screen is electrically-conducting metal which surrounds electrical equipment or cables. The metal strongly discourages electromagnetic (EM) waves from passing through it, either inwards or outwards. Shields block EM waves in two ways, either by reflection from or absorption into the metal or by both. Shields can vary in complexity, size and stiffness from a flexible wire braid round a cable to a steel-sheet-lined laboratory built to keep out EM disturbance. A simple tent built of metallized materials such as textile or wallpaper can also be a good electrically-conducting shield. Another example of a shield is the aluminium or steel hull of a yacht. A direction-finding radio is useless below deck surrounded by a metal hull because the completely shielded antenna cannot respond.

For a near-field magnetic source, absorption is the dominant shielding mechanism at all frequencies except the lowest. At frequencies as low as mains power (50 Hz in Europe, 60 Hz in the USA and on HM ships) magnetic fields can be countered by one of two methods. First, a highly ferromagnetic material can be used as a shield. Such materials are used around magnetic recording heads and some transformers (Fig. 9.1). Secondly, placing a conducting copper loop across the incident field induces from the magnetic field a current that conveniently creates a magnetic field in opposition to the incident field. In the near field, when one considers a magnetic source, the electric field can be neglected (Fig. 9.2).

The normal practice with a transformer reduces the leakage of low-frequency magnetic field by wrapping the magnetic source with a ring of copper strip and sometimes with two strips perpendicular to each other. Transformers create intense magnetic fields – this is how they

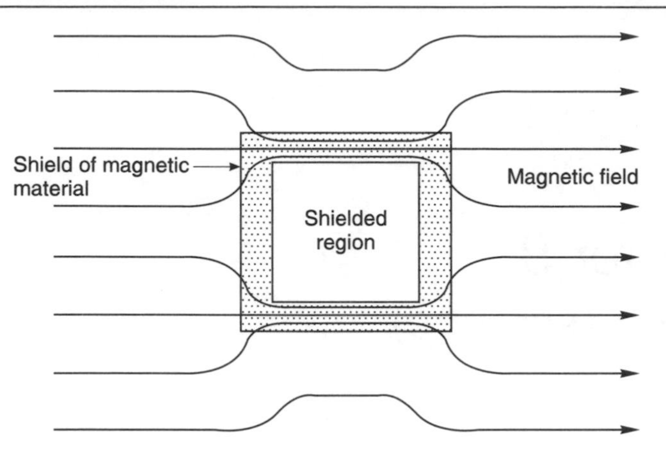

Fig. 9.1 Shielding with a magnetic material against a low-frequency magnetic field (after Ott, 1988)

work, but leakages from these intense fields can be prevented or greatly reduced by the encircling copper ring shown in Fig. 9.2, and often known as a 'shorted turn'. If the copper is perpendicular to the magnetic field, the field induces a current in the copper which opposes the magnetic field, greatly reducing it.

This method meets the strict requirements of the German VDE regulations against power-frequency magnetic fields. It thus would prevent disturbance to a cathode-ray tube which might be placed on top of the transformer. It is very difficult to shield people from such power-frequency magnetic fields although they can be shielded from electric fields.

Is a shield necessary?

A decision as to whether or not an appliance deserves the expense and trouble of building a shield round it has to be made early in its design. Deciding in favour of a shield is unpleasant because a shield makes the appliance less accessible, more expensive, heavier, slower both to build and to gain access for maintenance to the inside, also usually uglier. One criterion is if the EM interference emitted by the conductors in the printed-circuit board is calculated to be larger than permissible, a shield may be needed to prevent the appliance spreading this pollution. If the appliance creates interference, but of unknown origin, one may be forced to try shielding.

For a frequency above 1 kHz or an impedance above 1000 Ω, conductive shielding of copper or aluminium is usual. For a smaller

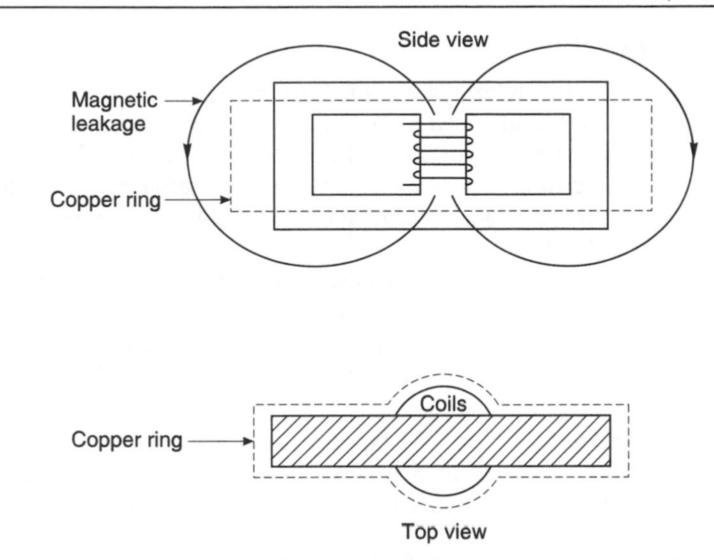

Fig. 9.2 Copper loop (shorted turn) to reduce the magnetic field emitted from a power transformer (after Paul, 1992)

impedance or a frequency below the very low frequency of 1 kHz, magnetic shielding of mu-metal may help, as in Fig. 9.1. But at frequencies between about 10 kHz and 1 MHz, steel is the most suitable and usually the cheapest material. At frequencies above 1 MHz copper may be better.

Ferromagnetic materials such as mu-metal have two weaknesses apart from their high cost. At frequencies above the lowest they lose their ferromagnetism and they also saturate quickly. To counter these weaknesses, several authors propose shielding with combinations of substances, including laminates, of different ferromagnetisms, with the lowest ferromagnetism on the side nearest the source.

The strong field thus first reaches the layer of lowest ferromagnetism. This substance saturates less easily than highly ferromagnetic substances. When the field, thus weakened, traverses the first layer of the shield and reaches the highly ferromagnetic layer, this layer does not saturate but further weakens the reduced field. The metal layers should be separated by gaps of air or other insulator at least as thick as the metal so as to avoid interaction between the fields in neighbouring layers.

For obvious reasons, a shield designed to prevent disturbances leaking out from an appliance will ordinarily be in the near field, while one to protect against incoming disturbance will usually be in the far field. Most shields are connected to ground or to the appliance frame, usually called the system reference (more on grounding in Chapter 12).

Wave impedance

Electric and magnetic fields need some more explanation. An electric field (E field) is found between two conductors at different potentials and is proportional to the potential difference divided by the distance between the conductors. A typical electric field is between the plates of a charged capacitor. E fields are measured in volts or microvolts per metre (V/m or μV/m).

A magnetic field (H field) encircles any wire or other conductor carrying a current, especially an inductor (Chapter 1, Fig. 1.3). It is measured in amps per metre (A/m). The two fields in an EM wave move together but different methods have to be used to shield against each.

The wave impedance of an EM wave, shown in Fig. 5.4, Chapter 5, is important to the designers of protective shields. The wave impedance is the electric field in V/m divided by the magnetic field in A/m, thus dimensionally:

$$\frac{V}{m} \div \frac{A}{m} = \frac{V}{A}$$

Consequently it is in ohms since, by Ohm's law, V/A = ohms (Ω). The diagram shows that in the far field the wave impedance is constant at 377 Ω. But in the near field it varies with its source, its distance from the source and the frequency.

Electric fields are easier to guard against than magnetic fields but the copper or aluminium used for protection against electric fields provides no magnetic shielding. The need for magnetic shielding is greatly reduced by ensuring that the conducting loops in circuits are small. This can sometimes be done by a two-sided printed-circuit board that allows the use of a ground plane (a copper sheet) on its underside. The ground connections made to the ground plane can be much shorter than when made to the ground, which naturally is further away.

Any conductor entering a shield should be connected to a filter to reduce or eliminate conducted interference. Some filters are small enough to be built into a connecting plug. For filters, see Chapter 11.

The AC skin effect: skin depth

Unlike DC, which flows uniformly over the whole cross section of wire, AC travels along the outside of its wire. With rising frequency the outward trend becomes stronger. This 'skin effect' greatly increases the resistance of the wire. The skin depth for a particular frequency of AC with a particular metal is taken as the depth from the wire surface at which the current density drops to 37% of its value at the surface.

Table 9.1 Table of skin depths for copper

Frequency	Skin depth (mm)
60 Hz (mains in USA)	8.5
1000 Hz	2.1
10 kHz	0.66
100 kHz	0.21
1 MHz	0.06
10 MHz	0.02
100 MHz	0.007
1000 MHz	0.002

With acknowledgement to Clayton R. Paul, 1992.

Thus, for high frequencies, thick wires are not more useful than thin wires. Generally the thickness of a shield should be more than the skin depth. Table 9.1 shows how, for copper, the skin depths diminish with rising frequency.

Shielding effectiveness

To the lay person the terms used to describe shielding effectiveness can seem contradictory so here is a short explanation. Good protection by a shield is described as high shielding effectiveness, and this means high loss of power by the incident wave, involving high losses from reflection and absorption. Most well-designed shields are thicker than the skin depth for the shield metal at the frequency of the incident wave. Compared with a shield of thickness equal only to the skin depth, the absorption loss doubles when the shield thickness doubles. At low frequencies, below 20 kHz, reflection loss helps the most for both ferrous and nonferrous shields. At higher frequencies ferrous materials have good absorption and shielding effectiveness. Generally a shield one skin depth thick provides about 9 dB of shielding effectiveness.

Slot effect

Most shields unfortunately need to be penetrated and great care is needed in the design of holes to allow cables to enter without destroying the shield's protection. Small holes are preferable to large ones because a slot with a length approaching a quarter of the wavelength in question can act as an antenna and broadcast most of the interference energy collected by the shield. So it is usual to limit slot lengths to one-hundredth of the wavelength at the highest frequency.

For practical purposes the effectiveness of a shield is decided by the holes through it, so to achieve a maximum shielding effectiveness of 20 dB, with frequencies up to 1 GHz, no hole should be larger than 16 mm across. At 1 GHz the wavelength is 300 mm (1 ft) and 16 mm is about onc-twentieth of a wavelength.

If the conductors inside the shield are all in the same plane, a small reduction of the loss of protection from a slot can be achieved by ensuring that the slot is aligned parallel with these conductors. Seams joined with spring fingers are looked on as a series of slots. A usually-acceptable alternative to a slot is a number of small holes, not more than 20 mm in diameter. If, however, any conductors run past the openings outside a shield, an appreciable worsening of the shielding effectiveness can be expected. Cables should therefore not be located near shield openings.

Backshell connector

From the above it should be clear that the holes and joints in a shield are its most important part. At any cable entry the cable shield should be connected for the full 360° to the appliance shield. This can best be done by a suitable backshell connector, which is expensive but it does the job! Any pigtail connecting the braid to the appliance shield should be forbidden.

Gaskets

A gasket is a conducting washer inserted into the gap between, for example, a door in a shield and its door frame. It makes sure that the gap is closed and thus eliminates any possible slot effect. Gaskets have to be electrically conducting and the best known types, made of knitted wire mesh, may be built into a PCB. Where a gasket is intended to prevent leakage at a door, it should be placed inside any fixing screws. If placed outside, it will not protect against radiation through the screw holes.

Gaskets maintain the continuity of a shield at a point where it must be opened. Some gaskets can be reused. Other types have to be replaced whenever the shield is opened. Apart from knitted wire, many other materials are used for gaskets, including silicone rubber loaded with electrically-conductive metal or with metal-plated particles. Loaded silicones should be handled with clean cotton gloves to avoid any possible corrosion of the included metal.

Gaskets ensure a continuous electrical contact, closing the hairline gaps that can ruin shielding at any frequency, especially when a quarter-wavelength is shorter than the slot length. They eliminate the high cost of accurate machining of a box and its lid or of a door and its frame, which would be needed without a gasket.

A groove designed to accept a gasket in the frame under the lid reduces the amount the gasket must be compressed to make the seal and ensures that it is correctly positioned. The commonest knitted gaskets consist of wire about 0.1 mm diameter made of Monel metal, copper, tinned copper, tinplated copper-clad steel, etc., but wire diameters range from 0.05–0.15 mm.

Thin-film shields

Where protection only against high frequencies of 30 MHz or more is needed, the shield for a box made of plastic may consist of a thin conductive coating over the plastic, but where exclusion of lower frequencies is also needed, solid metal is better.

Thin-film shields are those which are not self-supporting. They include metal foils, some of which may be stiffened with one or more plastic or paper sheets. They provide negligible absorption but can be excellent reflectors, especially at high frequencies, yielding a protection of 100 dB at 1 GHz or 60 dB at 1 MHz. Adhesive mu-metal foil is the most expensive but the most effective for shielding against low-frequency magnetic fields.

Shield adequacy

The usefulness of a shield depends on three effects added together: reflection from its outer surface, followed by re-reflection from its inner surface and absorption within the metal. Copper and aluminium, with their high electrical conductivity, reflect electric fields more efficiently than less conducting metals like steel. Re-reflection is important in thin shields at low frequencies.

Absorption depends on the shield thickness and increases rapidly with the frequency. All types of field are absorbed equally, whether electric, magnetic or plane waves. For magnetic fields, steel provides better absorption than copper of the same thickness, although excellent conductors like copper and aluminium provide better reflection.

No shield is perfect. Some interference will get through. Excellent shielding of 100 dB is sometimes possible. This means that the interference getting through is 100 dB less than without a shield. Such a high figure is not always needed. All shields against electric fields have to be electrical conductors but not all have to be solid metal.

Conductive plastics

Plastics, especially made to be electrically conducting by including in their manufacture a conductive filler of metal powder or flake or graphite or tiny metal-coated glass spheres, can be used for building a moulded appliance case but the addition of the conductive filler alters the mechanical properties of the plastic and cannot provide a conductive outer surface. Adding a conductive filler to a plastic is sometimes preferred to a thin film applied later because it is cheaper but it is not always so effective. Such plastic units must be given conductive metal connections round their perimeter, in the form of copper strips or multiple copper wires buried in the plastic so as to enable them to be electrically bonded to nearby units.

Some filler materials are expensive. Other constraints are that a high loading of metal (sometimes 80%) can cause severe tool wear, as well as reducing the quality of the surface finish and the strength of the plastic moulding. Because the distribution of the conductive material can be upset by the flow of the melt during processing, the moulding may be weakened and shielding may be lost, especially at corners or other changes of cross section.

Conductive paints for shielding

Electrically-conductive paints contain conducting powders in the same way as the conductive plastics mentioned above. They can be applied in different ways, including spray painting. High-quality modern paints are advisable so as to avoid any chance of loss of adhesion and flakes of paint dropping off, which has happened with older paints leading to the disaster of a slot effect already mentioned in connection with large holes through a shield. Lost flakes of conductive paint may cause another mishap, the short-circuiting of a printed-circuit board. Modern paints generally stick well but all plastic surfaces need to be cleaned and preferably wire-brushed before painting, to be free of oil, grease or other contaminants.

The use of a protective paint on a building wall is recommended by W.-D. Rose in his book *Ich Stehe unter Strom* (1996) especially to prevent a magnetic field passing through masonry into a bedroom where it could disturb a sleeper. He also recommends the use of mu-metal foil for the same purpose and considers it is equivalent to steel plate several mm thick.

Silver is by far the most electrically-conducting metal but it is also the most expensive one used for this purpose. Silver particles are usually held in a vinyl or polyurethane paint. Nickel paints are probably the next best after silver because nickel has good magnetic properties as well as conductivity. Nickel particles are usually held in a two-part acrylic paint system. Copper for various reasons has not been generally accepted. All these paints are touch-dry after half an hour and they dry sufficiently within an hour to take another coat, but generally the full electrical conductivity of the coat is not reached until much later. For nickel in acrylic as much as five days are needed for this stage to be reached.

A primer may have to be put on the plastic before the paint so as to improve adhesion. Specialist advice from the paint contractor may be essential on this point and on others.

Painting on metal

The paints mentioned do not stick well to metals. Nevertheless conductive paints may have to be put on to metal surfaces to prevent corrosion and to ensure good electrical contact between a gasket and the metal surfaces. Epoxy paints stick well to metal and, to make it conductive, the epoxy can be loaded with silver or with nickel. The nickel-epoxy must be applied only after the surface has been cleaned. Polyvinylidene fluoride or glass-reinforced units also can be coated with epoxy conductive paint. Nickel-epoxy paints are usually available in three parts, the base, the hardener and the thinner. They have good impact resistance and stick well to aluminium, steel or chromated aluminium. Nickel paints often have up to 80% nickel but higher percentages spoil the paint.

Metal-surfaced plastic or glass particles are preferred to solid metal particles from several viewpoints. They are cheaper, requiring less of the expensive metal and applying it more effectively than as solid metal. The AC skin effect forces the current to flow over the surface of the particles. Nickel-plated graphite has been used as well as silver plating over solid ceramic or tiny hollow glass spheres of about 50 microns diameter. Hollow particles have real advantages. Being less dense than pure metal they are held in better suspension in the paint, which needs less stirring. They do not settle out in a pump hose or, even worse, on a vertical painted area, leaving some parts unshielded.

Electroless plating

Electroless plating with nickel or copper is a chemical process in which the object to be plated is completely immersed in successive baths of chemical. The baths provide a coating of the metal both inside and outside the plastic unit. Cooperation between the designer of the plastic and the painting company helps to reduce cost and provide a better job by eliminating unplatable corners. The double metal coating provides a good shield and can be used even for large units, especially if manufactured in quantity.

Flame spraying of metal on plastics

Flame spraying of zinc is well established and is a much more effective shield than conductive paint but it is not an easy technique and it requires a skilled operator. An electric arc between zinc electrodes melts the metal, which is blown on to the plastic unit with an air jet. Unfortunately the heat can embrittle or distort the plastic because the hot sprayed metal is relatively thick, although this does provide a good shield. As with electroless plating, cooperation between the plastic designer and the flame sprayer can reduce cost and complete a better job. Flame-sprayed zinc does not corrode and has good conductivity. In general, flame-sprayed metal, because of its porosity, has a conductivity only about 10% of that for the same metal rolled or cast. But it is much better than metal-loaded paint.

Adhesive foil or tape

Adhesive copper or aluminium foil can provide shielding at plastic surfaces or on cables. The strongest foil is sandwiched between two layers of plastic or paper. It can be put on by hand but since no mechanical method of application has been found, it is more suitable for EMC laboratory work than for factory production. It is usually put inside the plastic box. Not every adhesive is conducting so it is possible for a layer of tape to be insulated from its neighbours.

Conductive caulks, glues and greases

Electrically-conducting glues, caulks and greases contain metal particles, just like the paints. Caulks are jointing substances that do not harden but maintain a permanent flexible seal. When a caulked joint is opened the old caulk is scraped off and the joint is remade with fresh caulk. Metal surfaces must, as usual, be clean, degreased and free of oxide. The caulks can be used on plastics or painted metal. Caulk is applied with a caulking gun or syringe or by hand with a putty knife.

Low electrical resistance in a switch or circuit breaker can be achieved by a conductive silver–silicone grease which also reduces arcing and its accompanying interference, as well as wear at contact surfaces. Usually it is stiff enough to be applied overhead without annoying drips.

Conducting glues are based on epoxy or silicone and contain silver particles. They hold gaskets in place in the absence of a slot, can fill bolt holes or screw holes and can make an electrical connection to metallized conducting textile, or bond together two sections of waveguide. As usual the mating surfaces must be well cleaned before the glue is applied. Epoxy glues even without silver are expensive but they are rapid-hardening enough to be used for bridge repairs. The silver in these substances is expensive but the bulk of the glue used is not large so the cost is not exorbitant.

Holes for meters or ventilation

To guard against the loss of shielding effectiveness caused by a hole that must be large enough for a meter or fan to operate through a shield, at least two solutions exist. An extra shield can be provided in front of a meter, in the form of a very thin silver or gold coating on the glass over the dial. It reduces visibility slightly but gives the glass a conductive surface that rejects electric fields by reflection. Conductive silver-impregnated glass can be used instead.

Another type of window that provides shielding for cathode-ray tubes or other displays includes a conductive wire-mesh screen laminated between two clear glass or plastic sheets. These screen-protected windows can have a transparency of up to 98% which is appreciably more than can be achieved by precious metal coatings.

Alternatively, behind the meter, an additional subshield can be added, with a connected capacitor. In any case the meter itself must not broadcast interference. This may be prevented by connecting a filter to the circuit supplying the visible display. The subshield behind the meter must make firm, all-round electrical contact with the main shield. Merely covering an opening in a shield with a metal screen does not close the opening. To be effective, the screen must be continuously bonded around its edges. If this bond is absent the opening has the same disastrous slot effect as before.

Honeycomb waveguide

At high frequencies, for a large opening like a ventilation or control shaft, there is only one shielding solution. The opening is built with a metal honeycomb waveguide inside, acting as a high-pass filter. Any waveguide has its own cutoff frequency below which there is considerable reduction of frequencies transmitted. High frequencies above the cutoff frequency are transmitted without reduction. The length of the honeycombs determines the amount of reduction while their diameter determines the cutoff frequency. If the honeycomb cells are at least four times as long as their diameter, the shielding effectiveness can be expected to reach the excellent level of 100 dB.

The honeycomb must not be designed as an air cleaner and so in addition to the honeycomb, the opening may be provided with a filter to clean the air. It consists of wire mesh, sometimes coated with oil to catch dust, which must be periodically cleaned, either by an air jet or by washing in a solvent that removes the dirty oil. The filter is built into an aluminium channel frame that is easily removed from the honeycomb.

Specialized shielding materials

Metallized conductive textiles may be made of many substances including polypropylene, nylon, polyester, aramid, glass, carbon fibre, silicone or alumina. The fibres are typically 10–250 microns in diameter, with a thin metal layer 0.1–5 microns thick to provide some conductivity. Some of them are partly transparent, all are lightweight and metallizing does not alter their shape, shrinkage, elongation or strength.

These textiles can be joined by riveting, stapling, and gluing. Some of them can even be soldered with conventional 60–40 tin–lead solder. Though not so transparent as some ultra-thin metal films, metallized textiles can have open areas up to 50% for shielding windows, VDU screens, ventilation openings, etc. Metallized textiles can be made watertight by coating one or both faces with a film such as polyurethane, polyvinyl chloride, silicone or neoprene. These textiles can make good gaskets for a waveguide. W.-D. Rose in *Ich stehe unter strom* (1996) recommends a textile of 88% cotton interwoven with 12% of very fine stainless steel wire, that can be cut with scissors and used as curtaining or for clothes.

Electrically-conducting wallpaper

Electrically-conducting wallpaper can provide almost as much protection from interference as the more expensive, cumbersome and heavier metal shields. Usually nickel-plated carbon fibres are randomly crossed in a nonwoven fabric resembling a mat made of fine hair.

The lightweight flexible wallpaper can be glued down with ordinary nonconducting wallpaper paste. Overlaps are needed but do not have to be covered with conducting material because of the random orientation of the nickel-plated carbon fibres in the wallpaper. It can be joined to a door frame or other metal unit by soldering without loss of mechanical properties. It is used for shielding anechoic rooms, or for protecting civil and military vehicles including the electronics of aircraft.

In August 1996, an article in the Russian newspaper *Izvestiya* mentioned a Russian metallized textile known as Voskhod which is elegant enough to be perfectly suitable for city clothes on men or women and can be recommended for wear by people who need to be shielded during their work at computers or similar devices imported from the west. The report states that western safety limits are a thousand times more unsafe than Russian limits (as explained in Chapter 14), consequently shielding from such devices is desirable. It is effective both at low and at high frequencies.

Cables

Because they are long and can act as antennas, cables can bring more interference into an appliance than any other component, so they and their end connections must be carefully specified. Screened cables may be needed and screens can be provided during manufacture.

Different sorts of conductors are used especially for signals at different frequencies. Up to 100 kHz including DC, the twisted-pair conductors familiar to telephone engineers and usually unshielded, are commonest. Coaxial cables also can be used for DC but are usual for any radio frequency current up to 100 MHz, while above 1 GHz, waveguides are the rule.

Cables may sometimes not need screening, if provided with an effective filter at the junction with the appliance which they serve. Filters can remove hum and noise from a mains power conductor and are dealt with more fully in Chapter 11. Conductors carrying high-frequency currents should be separated from others in which they could cause interference. The distinction between 'high' and 'low' frequency depends on circumstances and may be only a few kilohertz.

For house wiring, mild steel conduit encasement to cables became out of date during the 1960s and was replaced with the very much more convenient, cheaper, and less bulky, plastic-insulated cables. Where the electromagnetic field emitted by cables is too high, however, steel conduit may have to be used although it is expensive to install and bulky. It has excellent shielding power if it is grounded (earthed).

Metal braid or flexible-metal conduit may be enough to shield a cable. The braid is woven from copper or tin-plated copper-covered steel wire and various grades of braid exist, with coverage of 60 to 95%, the latter being the best. As ferromagnetic materials (iron, steel, mu-metal) make good shields against low-frequency magnetic fields but have a much lower electrical conductivity than copper or aluminium, a combination of several metals is recommended by Degauque and

Hamelin (1993). They propose as many as three steel braids, two of them covered by spirally wound ferromagnetic tapes such as mu-metal.

At higher frequencies, as the wavelength shortens, the gaps between the braid links become a larger proportion of the wavelength, involving a greater likelihood of a slot effect. Multiple shields of braid improve matters but are expensive and less flexible than a single braid. Solid aluminium foil shields for cables exist but are weaker than braid and difficult or impossible to terminate properly. Degauque and Hamelin propose aluminium foil wrapped round the braid so that the braid can be correctly terminated and the foil covers most of the holes in the braid.

Ferromagnetic materials

Mu-metal has excellent magnetic properties but like other ferromagnetic materials is not always simple to handle. After manufacture any mu-metal object is formed by annealing in a strong magnetic field in hydrogen to align the magnetic domains and create the high magnetic properties. But if the unit is dropped, drilled or bent, the magnetism may be lost, although it can sometimes be recovered by annealing. It is thus difficult to make a mu-metal unit that is large and, furthermore, because mu-metal is easily saturated it cannot be used in an intense magnetic field. All such highly magnetic materials lose their magnetism with rising frequency. Where their magnetic properties are important, these materials should not be allowed to carry intense direct currents which could saturate them.

Makers of ferromagnetic materials quote their ferromagnetic properties at 1 kHz. They choose this very low frequency because it gives the highest possible value but at the relatively low frequency of 20 kHz the ferromagnetism of mu-metal drops to that of cold rolled steel. These expensive ferromagnetic materials are thus suitable mainly for magnetic fields below 1 kHz and for higher frequencies steel is equally effective and much cheaper.

Shielding against power-supply frequencies of 50 or 60 Hz can be effective with mu-metal if it is not saturated. The heavy currents in power supplies could, however, saturate the mu-metal.

A cable from a computer to one of its peripherals is likely to be 1.5 m long and this length, a quarter-wavelength at 50 MHz, could make the cable into an antenna for this frequency, which unintentionally radiates interference. Shielding the cable might help.

Ribbon cables

Ribbon cables are often used for data transmission in buildings because they can easily be laid under a carpet. Shielding of a ribbon cable is possible but the shield is difficult to terminate against the shield enclosing the appliance. The layout of the ground conductors in a ribbon cable seriously affects the elimination of interference. One good arrangement is to have a ground conductor next to each signal conductor. It is cheaper to use only one ground conductor but this creates large polluting inductive loops with resulting interference from the signal wires that are furthest from the ground wire. Ribbon cable also exists in which there is a continuous copper ground plane underneath the conductors, which may be the best arrangement but it is awkward to work with and expensive.

Buildings

Users of mobile phones know that buildings have an unwanted shielding effect on radio waves which varies with the frequency. At 100 kHz in a reinforced concrete building the reduction in strength of EM waves is about 15 dB for the magnetic field and 50 dB for the electric field. At frequencies above 1 MHz both fields are reduced equally by about 20 dB and at or above 1 GHz (1000 MHz) passage through openings becomes easier.

Chapter 10
Transients and protection against them

Transients are single events, short pulses (surges) of extra high or extra low voltage or current. On a current or voltage graph, a spike is a steep transient with a pointed summit which generally lasts some 10 milliseconds. Lightning is probably the most powerful transient, with many thousands of volts and hundreds of amps amounting to a power of 1000 kW or more. A lightning strike is extremely short and it occupies usually only 20–30 microsecondss and rises to its maximum in a fraction of a microsecond. All the lightning flashes of one event can occupy 0.2 s. Direct lightning strikes bring transients to both common-mode and differential-mode currents. See Chapter 7 (also Chapter 15, Glossary) for an explanation of common mode and differential mode.

Chapter 3, Power supplies, deals with voltage regulation, a subject related to transients but with a different emphasis. Power-supply voltages have to be kept steady which of course is a wider subject than dealing with transients alone. Suppression of transient surges must be fast, usually within nanoseconds, but for the voltage regulation of a power supply, such speed is less important.

Man-made transients are usually very much smaller than lightning spikes although much depends on how near the lightning is. Electrostatic discharge (ESD), described in Chapter 6, is a common source of man-made transients which also come from switching, especially of inductive loads, such as electric motors, generators or solenoids. All of these have bulky coils of wire with iron cores that are able to store considerable energy. Their starting or stopping peaks can reach 15 times the normal voltage, increasing to the maximum in only 0.01 μs. The worst peak comes on switching off, when the stored energy is released.

Many other types of switching transient exist, often picked up by the electric power cable and brought into the electronic appliance both as common-mode and as differential-mode noise.

There are several types of switching transient, of which the first occurs as soon as the power is turned on. For a conductor 10 m long the pulse may occupy several tens of nanoseconds. A few microseconds after switching on or off the 'late transients' occur. When this pulse meets any change of impedance, part of the pulse is reflected, leading to complex waveforms oscillating up and down the cable in periods of much less than a microsecond.

Protection against transients

Surge protection devices operate when a certain voltage threshold is exceeded. They function like a switch that offers a short-circuit path to the transient. The impedance of the short circuit is arranged to be very much smaller than that of the parallel path through the circuit being protected, so as to provide an 'incentive' that diverts the transient to a harmless path.

Protection against transients is obtained by cutting the voltage to a safer level but the protection achieved depends both on the amount of voltage reduction and on the speed of operation. Since transients can occur in nanoseconds, protection also must function within nanoseconds. Filters cannot handle the large amplitudes of switching and lightning surges, so surge protection must always be installed 'upstream' of any filter. For lightning protection 'surge diverter' is the accepted general term, 'surge suppressor' being less exact although they are also called 'terminal protection devices'. Lightning is too powerful to bc absorbed or arrested, it must be diverted.

Most modern surge suppressors are electronic and often several of these devices are used together and they include:

- semiconductor avalanche suppressors,
- metal-oxide varistors alias voltage-dependent resistors,
- gas-discharge surge diverters, (gas tubes),
- spark gaps, (lightning arrestors),
- power conditioners,
- expulsion-tube arrestors.

Circuit breakers are not used, they are much too slow for surge suppression.

The semiconductor avalanche suppressor (SAS) can deal in nanoseconds with a transient of 15 kW. The transient power is calculated as the peak volts times the peak amps. It is extremely fast (15 picoseconds (ps)) but cannot take a very powerful transient so is usually installed 'down stream' from other diverters, and protects signal or data lines rather than power lines. SAS also have a large capacitance, up to 15 000 picofarads (pF). Ratings of SAS vary from 1.5 to 15 kW.

The **varistor** is made of sintered metal oxide, usually zinc oxide and may be called a VDR (voltage-dependent resistor) or an MOV (metal oxide varistor). Its response time is fast to medium (25 ns) but it can handle either power or signal currents. Varistors have a steady demand for power, of the order of 2 W, but this usually does not matter because they are so versatile. Those that are mounted on a PCB are called surface-mount varistors. Larger ones handle 2800 V or more. Varistors have high capacitance of 200–2000 pF which can cause problems at high frequencies.

When the energy rating of a varistor is exceeded, it fails usually by short circuit or by blowing the system fuse and if the varistor is not fused it may be destroyed, ejecting solids and gases. A varistor protected by a fuse should also have a monitoring circuit to check whether the fuse has blown. The life of a VDR depends on how much current it has diverted. For instance a VDR which could last only one pulse of 100 A for 0.1 millisecond (ms) could last for a million pulses of 10 A, also for 0.1 millisecond.

The **gas tube** is a pair of electrodes in a gas at low pressure. When the transient voltage between the electrodes reaches the gas breakdown voltage, the transient is diverted to ground. The SAS and the varistor both absorb the transient, but the gas tube diverts it. Gas tubes protect both power lines and telecommunications lines, often in conjunction with a varistor or SAS. Gas tubes have a response time of 0.5 ms which is regarded as slow but time is needed to ionize the gas. They have a small capacitance, less than 10 pF, so work well at high frequency.

The **spark gap** (lightning diverter) was originally used both for power and for telecommunications before the gas tube superseded it. The transient is diverted to a ground electrode. Early spark gaps suffered from power-follow current, which continued to flow at the normal voltage after the transient had passed because the spark gap's impedance had been lowered by the ionization of the air caused by the transient. Several gaps can be placed in series to increase the breakdown voltage. To limit power-follow current a resistor such as silicon carbide is placed in series with the gap as a so-called 'valve'. It has a low resistance under high voltages and a high resistance under low voltages. The best valves quench the power-follow current within one halfcycle of the power frequency (in about 10 ms) but, compared with the old spark gap which works in air, the gas tube is better with its low-pressure inert gas surround.

Power conditioners (PC) have no batteries, unlike uninterruptible power supplies (UPS). They can, however, withstand a power blackout of a few milliseconds and are effective against spikes, under-voltages, and oscillatory disturbances on the line. An expensive PC has a ferro-resonant transformer to compensate for input line fluctuations and transient suppression.

Expulsion-tube arrestors have a set of electrodes in a vented chamber. Around the electrodes in disk form is a substance that emits gas under arc discharge. The expansion of the gas, caused by the heat of the arc, blows out the arc. The impulse breakdown value of the arrestor is stated in volts or kilovolts per microsecond. It can operate at up to one million volts per microsecond, the level of an electromagnetic pulse, but it is not fast – other devices act more quickly.

Hybrid devices are combinations of two or more suppressors. The combination improves the speed or handling capacity in terms of voltage, current or energy. In one hybrid arrangement the first suppressor 'clips' the voltage and dissipates much of the energy and the second suppressor is fast and clips the voltage further. Gas tubes, SAS and varistors in different layouts make excellent hybrid devices.

Design of protection

The energy of the transient must be dissipated or diverted to bring the voltage to a level low enough to prevent damage. The maximum tolerable overvoltage must be estimated for the appliance concerned, but this must be done in conjunction with an estimate of risk. What degree of risk is allowed?

1. no upset,
2. temporary upset involving restart of the system,
3. damage to the appliance.

Condition (1) is not possible without astronomical expense and normally some data are corrupted. There is of course always a risk that the transient will be larger than expected and it may then destroy not only the transient suppressors but also all the equipment they should protect.

Flashover

Lightning flashover can mean a discharge from one cloud to another, which need not concern us here. But flashover at or near ground level may cause damage and any impedance to ground should be either low or negligible. Flashover happens when a badly installed ground conductor or lightning conductor of high impedance carries a rapidly changing current which arcs over to other metal. The impedance of the ground conductor has three components: the inductance of the wire, its ohmic resistance, and the resistance at the ground rod between it and the soil. The inductance of the grounding wire is the most significant part of the impedance, especially with lightning. It must be minimized by using the shortest possible path to the ground rod.

If the ground conductor or lightning conductor passes near metal such as a water pipe, cable conduit or building frame, an arc may flash from the lightning conductor over to the metal. The arc naturally produces EMI both broadcast and conducted. Where a ground rod might be prohibitively expensive the ground conductor may sometimes be connected to nearby metal objects, provided the connection is well bonded.

Installation of transient suppressors

All wires to suppressors should be as short as possible to keep their impedance to the minimum. Transients induced by lightning are common-mode currents and their very fast rise time allows significant voltage to build up on the ground conductor if its impedance is not extremely low. High impedance on the grounding conductor can thus allow some of the energy of the lightning into the protected equipment and may result in damage to the appliance or arcing by flashover. To keep the lengths of suppressor conductors as short as possible they are sometimes eliminated, for example, by the installation of the suppressor on a PCB.

Chapter 11
Filters

A filter is an electrical device or subcircuit that improves the quality of the frequencies wanted by decreasing the less desirable ones. The frequencies that it allows to pass are called the passband, those which it blocks are the stopband and the edge of the passband separating it from the stopband is the cutoff frequency.

A heavy current at a very low frequency of 50 or 60 Hz needs quite different treatment, involving a heavy, expensive filter, compared with the small current needed for communication or signalling at any higher frequency. Violette, White and Violette (1987) go into greater detail than is possible in this short book by describing the different filters needed.

Power-line filters are almost invariably lowpass filters. All the power cables, with their conducting shields, as well as any unshielded wires in the building can pick up currents at unwanted frequencies. As nearly all interference and noise are at much higher frequencies than mains power at 50 or 60 Hz, lowpass filters efficiently exclude them.

Filters are one way of eliminating unwanted frequencies that carry interference or noise. Other methods, more elaborate, expensive and often more effective than filters, are mentioned in Chapter 3, Power supplies. A simple one is the common-mode choke or ferrite ring described in Fig. 3.1 of that chapter.

Filters can be extremely lightweight (a few grams) although, as they are normally contained in a relatively heavy metal shield, such low weights are unusual. The size and weight of a filter increase with the rated voltage and current.

The workings of some common filters are shown in Figs 11.1(a)–(d). These four types of filter, although extremely simple, need to be

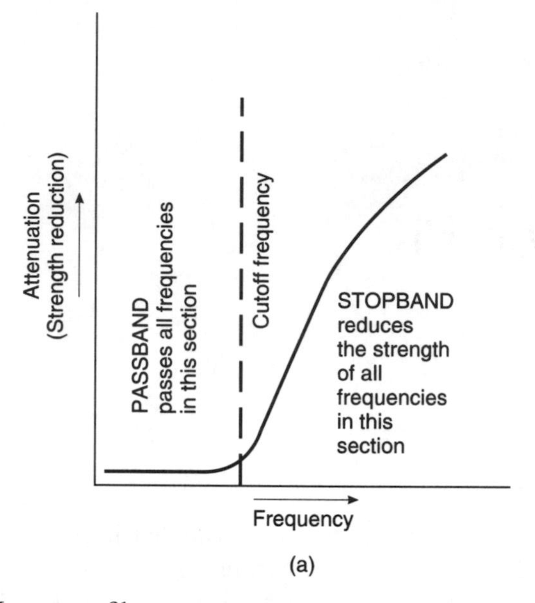

Fig. 11.1(a) Lowpass filter

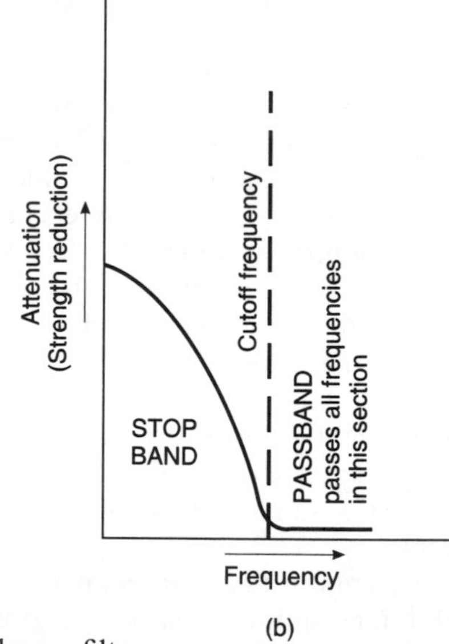

Fig. 11.1(b) Highpass filter

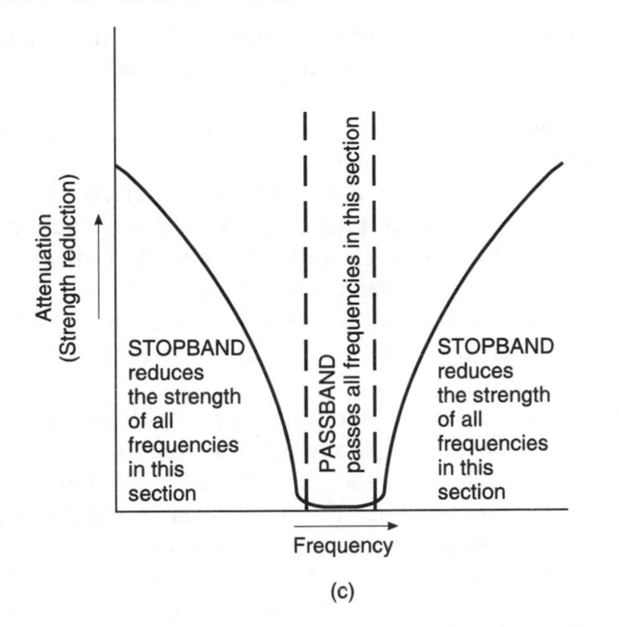

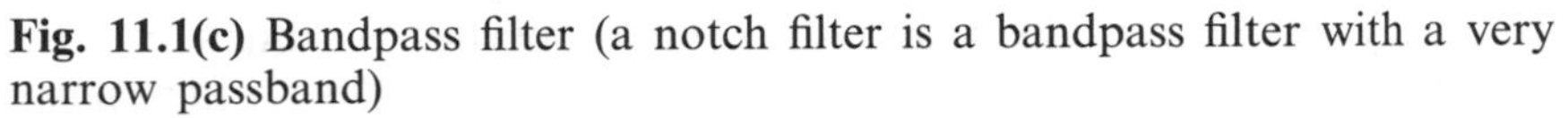
Fig. 11.1(c) Bandpass filter (a notch filter is a bandpass filter with a very narrow passband)

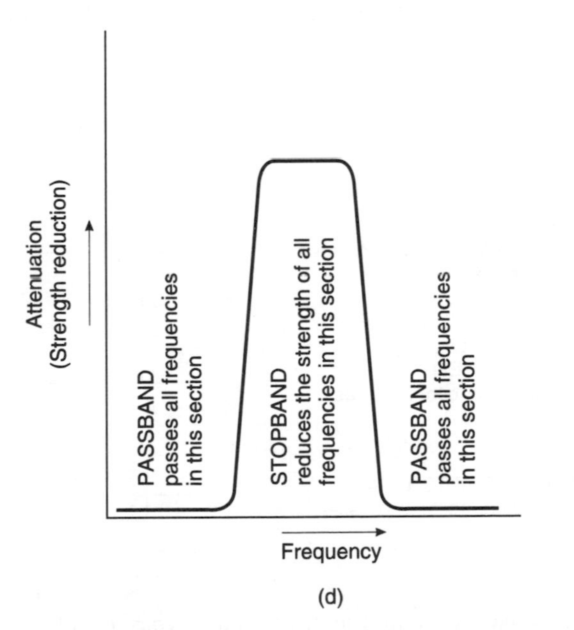

Fig. 11.1(d) Bandreject filter (with acknowledgement to Geoffrey Shorter)

designed by an electrical engineer to ensure that the reactances of capacitor and inductor suit the frequencies concerned.

- The lowpass filter needs nothing but an inductor in series with the circuit (Figs 11.1(a) and 11.2(a)).
- The highpass filter needs only a capacitor (Figs 11.1(b) and 11.2(b)).
- The bandpass filter needs both an inductor and a capacitor in series with the circuit (Figs 11.1(c) and 11.2(c)). A notch filter is a bandpass filter with a very narrow passband.
- The bandreject or bandstop filter needs a capacitor and an inductor in parallel with each other on the circuit (Figs 11.1(d) and 11.2(d)).

Hi-fi enthusiasts know the specializations of loud speakers, the 'woofer' for the bass notes, the 'squawker' for the middle notes and the 'tweeter' for the top notes. On the electrical circuits of these speakers there is invariably a lowpass filter for the woofer, a bandpass filter for the squawker and a highpass filter for the tweeter.

In practice no filter is perfect. No filter allows the passband through in all perfection, nor does it quite block its stopband, and the cutoff frequency is never a clean division. The properties of filters can be verified in sophisticated communications receivers. They often have filters which can be switched in to reduce noise by restricting the bandwidth. Although these filters can destroy the quality of music they can nevertheless be useful at 40 m (7.5 MHz) if speech alone is needed.

(c) BANDPASS FILTER
Inductor and capacitor in series

(d) BANDREJECT FILTER
Inductor and capacitor in parallel

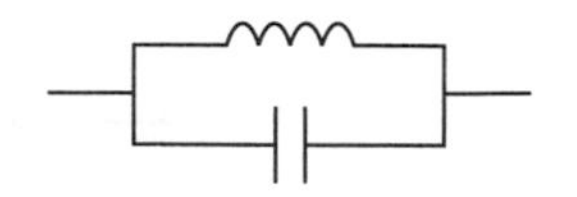

Fig. 11.2 Reactances used in the four main types of filter (after Paul, 1992)

If a source of interference is known, the filter to remove it should be placed as close as possible to the source. Especially at frequencies above the lowest, the input (dirty) wiring to the filter should be separated from the output (clean) wiring. If this is not done, the input current will pollute the output current and ruin all the good work done by the filter. Before choosing a filter, one must find out the maximum possible voltage, current and frequencies in the circuit to make sure that the filter can carry these maxima.

Insertion loss

All filters have an insertion loss. The insertion loss of a filter shows how much its presence reduces the current or voltage of the wanted frequency and is sometimes measured as a voltage reduction at the frequency of interest, but is usually stated in decibels. Typically it can be expressed as a function of frequency. The insertion loss of a particular filter depends on the impedances of source and of load and therefore cannot be stated independently of these impedances which will also vary with frequency.

Measurements of insertion loss for the (unwanted) common-mode (CM) current and the (wanted) differential-mode (DM) current are made by contrasting methods. DM current is always carried on two wires. By definition it flows out by the live wire and returns by the neutral wire, without using the ground (earth) wire. Consequently for DM current the live and neutral wires are the circuit to be tested and the ground wire is left untouched. To check the insertion loss for CM current the live and neutral wires are joined so as to form the supply circuit to the ground wire.

Jasper Goedbloed, who is in charge of EMC training at Philips headquarters in Holland, in his book (Oxford, 1992) affirms that a mains filter is a last resort. Ideally one should design the electronic unit to such a high standard that a mains filter is not needed.

Even if the right filter is connected it often does not satisfy expectations, possibly for some of the following reasons:

- long unscreened wiring between the protective ground (earth) and the mains filter;
- long wires between the 'clean' end of the filter and the circuit to be protected;
- inclusion in one cable bundle of both the 'dirty' cable to the filter and the 'clean' one from it.

These three mistakes are ways of inviting unwanted noisy energy into the 'clean circuit', whether by capacitance or induction.

Electrical connections to painted metal or anodized aluminium will not allow any current to flow.

The best results from a filter can be obtained when the filter is correctly surrounded by a metal housing. This is all the more essential if 50 dB or more of protection are needed at frequencies above 1 MHz.

Typical filters are either RC (resistor-capacitor) or LC (inductor-capacitor) types. The RC is 'dissipative', meaning that the noise is converted to heat in the resistor, although the LC filter is more effective at high frequencies for the same loss in mains voltage. However, the mains voltage drop with the RC filter may be worthwhile because, unlike the LC filter, it eliminates the noise permanently, losing it as heat. Both types have capacitors connected to ground (earth). With all filters the exposed (unscreened) wiring must be as short as possible.

One much overlooked but simple way of minimizing noise is to reduce the bandwidth of frequencies in the system to the absolute minimum required by the signal. Another help is to use slower logic in preference to fast logic. Fast logic generates extra interference and is also likely to be more susceptible to interference than slower logic.

Many electronic appliances contain an inductor connected to the green and yellow (green in the USA) safety wire. Its considerable impedance at the high frequencies of the interference currents effectively blocks them but it still allows the safety wire to be used for safety purposes – protection against electric shock.

A ferrite ring may help because there is no risk of this becoming unsoldered, as might happen with an inductor under a high transient current. The ferrite ring has to be chosen to be suitable for the frequency range in question. One typical value is 0.5 mH which gives an impedance of 1400 ohms at 450 kHz, the lower limit of the FCC. At the higher frequency of 30 MHz one could expect the impedance to rise considerably but this does not happen. Stray capacitances between the windings of the ferrite ring work in opposition to its inductive reactance and reduce the impedance.

Chapter 12
Printed-circuit boards: layout and grounding

Inside computers there are several levels of 'packaging', which it is helpful to understand. At the first, microscopic level, smaller than a millimetre is the IC, or integrated circuit, also called a chip or microchip. At the second level, several of these may be combined to form a chip package such as a DIL (dual-in-line) plug built with two rows of copper pins that connect to an appropriate computer socket. Originally there were seldom more than 20 pins. Now there can be many more. Somctimes it is not fitted with pins but soldered directly on to a printed circuit – surface mounted. The package that is surface mounted on a printed-circuit board (PCB) may also be called a componcnt or module or chip carrier. Printed circuits are the third level. They may be either cards, which are thin and smaller than a postcard, or boards, which are thick, often multilayer and sometimes carrying more than one card. At the fourth level up are metal frames or racks which carry several PCBs.

Planes of several types are important parts of two-layer or multilayer PCBs. They are sheets of continuous copper foil. A heat-sink plane accepts heat and removes it to an area where it can do no harm. Two other planes may also act as heat sinks. One of these, a ground plane, may be the second copper layer of a two-layer PCB. It acts as a common electrical reference point for ground (earth) connections and circuit returns. The second type, a power plane, also called a voltage plane, is one that is held at a voltage above ground potential and used in four-layer PCBs.

The PCB design involves combining such units into a whole which neither overheats nor disintegrates for any other reason.

EMC design is full of contradictions. Those who design computers or other electronic equipment naturally want their unit to work as fast as possible, as do people who buy them. This means fast switching, which for EMC is undesirable as shown in Fig. 1.5, Chapter 1. Good EMC design therefore involves discussion and cooperation between the mechanical engineer, the designer of the PCB and the electrical engineer, who should also be the EMC designer.

Printed-circuit boards

A PCB is an insulating board, usually of glass-fibre reinforced with epoxy resin, on which is fixed a layer of copper foil, of about 0.04 mm or even thinner. A circuit is photographically printed on to the copper and then most of the copper is etched away by acid. There may be a second copper sheet on the underside of the PCB, and multilayer PCBs also exist. After the etching is complete, a typical PCB has conductors called tracks, often only 0.5 mm wide. If it is 75 mm long its resistance will be 0.1 ohms and its impedance at low frequencies is likely to be 1 ohm. During its design stage, the resistance of this PCB track can be reduced by widening it or laying solder on the track or soldering a wire along it.

On a single-layer PCB, the tracks must often take roundabout routes that create high resistance. The use of a double-sided PCB with tracks on both faces means that the tracks can be made less circuitous and their resistance reduced, although this type of board is slightly more expensive.

Most PCBs contain chips which are so small (about 1 mm across) as to be almost two-dimensional. The current trend is to pack more and more circuits into each chip, increasing the number of layers in the IC. This can increase the effects of stray capacitance at high frequencies but where a PCB contains a number of chips, the high resistance of its narrow tracks can be helpful in reducing the currents between chips and thus may lessen their chances of picking up interference.

As usual, the reduction of lengths of wiring in circuits reduces stray capacitance and inductance and consequently interference. The PCB layout therefore has to be carefully designed, aiming at short conductors.

Because an IC chip is so tiny and only connected to the world outside its pins by a few fine wires, it is unlikely, even in a strong electromagnetic field, to pick up a large interference signal although the fine wires can certainly pick up something.

A chip is a complex unit with several circuits closely packed together, so internal interference can be picked up by one circuit from its neighbour, helped by stray capacitance or inductance at high frequency. If a chip has been mass produced and carefully tested, such internal interference is unlikely. It is more likely in a chip programmed for one particular user. Such special one-off chips are usually larger, up to 100 mm across, but it is often possible to design screening locally into such a chip to prevent pick-up of interference.

Grounding, avoiding loops

Grounding (earthing) has different meanings for different people, but because it is so important for the prevention of interference, several senses of the word will be dealt with.

The safety ground (the yellow and green wire in Europe, green in the USA) should have a low impedance to offer good protection against electric shock. In other words, any leakage of electrical power should be able to get away quickly to the earth without endangering people. But the ground wire should not be used as a return wire for signals. A signal circuit should have its own return wire.

Where RF disturbance from the ground is expected, and this is common, an exception can be made to the rule that the safety earth must have ultra-low impedance. One or more small RF chokes of about 1 mH (1 millihenry) can be inserted into the safety wiring. At 1 MHz this 1 mH choke has an impedance of about 6000 ohms, blocking disturbances at this and higher frequencies. At 50 or 60 Hz, the frequency of the mains, its impedance is very low, only 0.4 ohms or less.

As a protection against electrostatic discharge (ESD) every exposed metal part of the appliance must be grounded. Any breaks at joints, hinges, etc., must be made continuous with absolute certainty by using flexible bonding straps or another method to create electrical continuity across breaks. Any break in continuity in a metal enclosure to equipment can lead to electrostatic charges finding their way into the appliance's circuits and these are easily damaged by even the slightest ESD. Generally ESD grounding is multipoint so as to keep the impedance low and the wiring short.

Single-point or multipoint grounding?

At frequencies below 1 MHz, single-point grounding is usual although it can be cumbersome because it uses so much wire (Fig. 12.1). Above 10 MHz, the inductance of the long grounding wires rises astronomically so single-point grounds become impossible. Another disadvantage of long wires is that if any wire length coincides with a multiple of one quarter-wavelength it can act as an antenna and radiate interference.

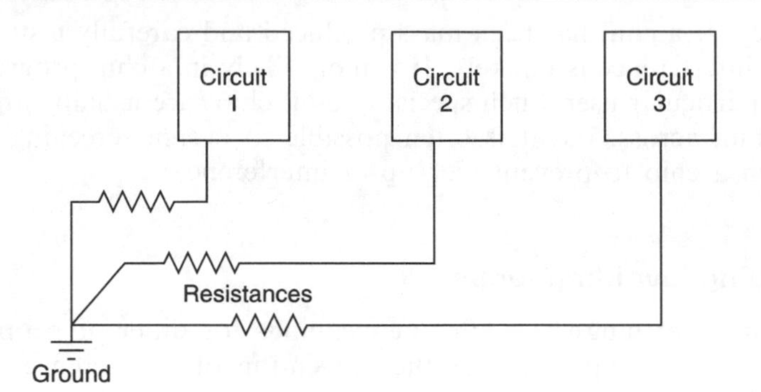

Fig. 12.1 Single-point grounding, suitable for low frequences, but often with awkwardly long wires (after Ott, 1988)

To insure against this possibility ground conductors should always be shorter than one-twentieth of a wavelength at the highest frequency.

At very high frequencies (VHF is 30–300 MHz) on appliances containing digital circuits, multipoint grounds can shorten the lengths of ground conductors enormously and minimize the impedance to ground. Each circuit is therefore connected to the nearest low-impedance ground available, usually the metal casing of the appliance (Fig. 12.2). In VHF work, ground leads have to be kept to a fraction of an inch because wavelengths are so short at VHF.

At high frequencies (HF is 3–30 MHz), with the aim of shortening the connections to ground and reducing their impedances, a ground plane of copper foil can be used to receive all the ground wires.

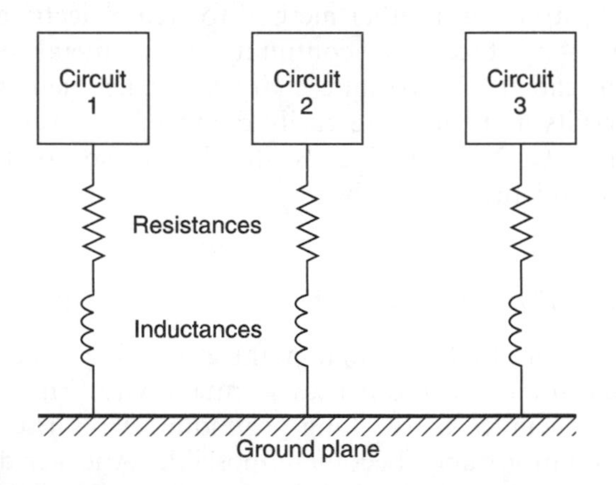

Fig. 12.2 Multipoint grounding, suitable for frequencies above 10 MHz if impedances to ground are minimized (after Ott, 1988)

Because of skin effect, the thickening of the ground plane does not reduce impedance at high frequencies. Silver-plating its surface, however, does reduce its impedance. The effect of the ground plane is to reduce the areas of loops and the likelihood of interference pick-up. The ground plane is usually the lower layer of a two-layer PCB.

At low frequencies (LF is 3–300 kHz) multipoint grounds are avoided, single-point grounds are usual. But at higher frequencies of 1–10 MHz a single-point ground is often possible provided that the longest ground conductor is shorter than one-twentieth of a wavelength. If it is longer, multipoint grounding must be used to reduce the length of wiring.

Almost as good as a ground plane in reducing the impedance to ground is a ground grid. If a ground plane is not acceptable because of the high cost of a multilayer PCB, a ground grid can be used. This is a two-sided PCB with vertical tracks on one side and horizontal tracks on the other at an appropriate spacing (about 12 mm) to suit the size of the units to be fitted on the PCB. Holes through the PCB can be made without difficulty to connect to the other side where needed. The ground grid is usually laid out before anything else on the PCB.

So far as computers or other digital PCBs are concerned, either a ground plane or a ground grid must be used. Both involve double-sided PCBs, occasionally multilayer PCBs.

Hybrid grounds

A hybrid ground is one that acts equally well although differently at different frequencies, changing its performance to suit the frequency (Fig. 12.3(a) (b)). Figure 12.3(a) works as a single-point ground at low frequency but as a multipoint at high frequency. Figure 12.3(b) is the opposite, multipoint at low frequency and single-point at high frequency. The inductors isolate the ground at high frequency. This relatively uncommon system is needed when several chassis must be grounded to the green and yellow wire for safety but where the signal circuits need a multipoint ground.

When different types of circuit exist on the same PCB they should be segregated so that all fast digital circuits are together and away from noisy circuits. Any analogue circuits also should be kept together and away from the others (Fig. 12.4). Each circuit should be grounded as appropriate to its frequency.

Grounding of metal racks carrying equipment

Large electronic systems are usually carried on metal frames or racks. In the old type of electromechanical telephone exchange the racks carry

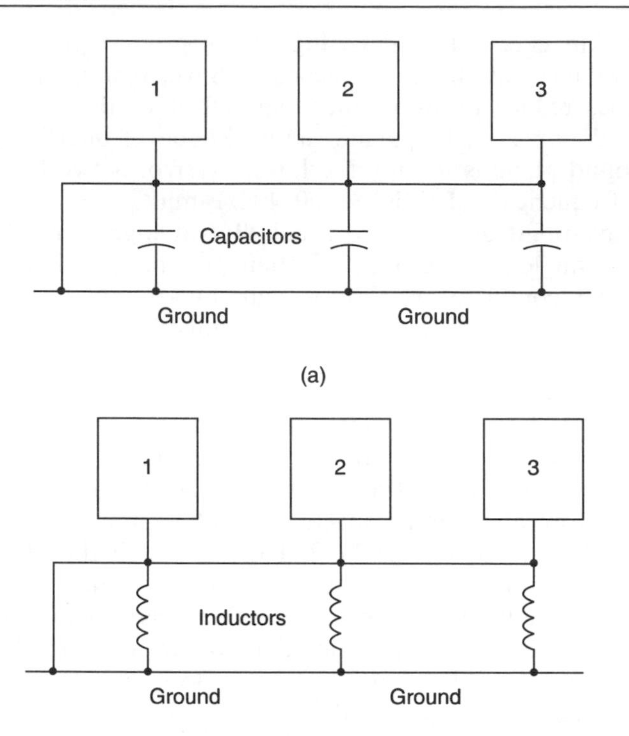

Fig. 12.3 (a) Hybrid connection to ground. At DC or low frequency the capacitors block any current flow and create a single-point ground. At high frequency the capacitors allow AC to flow and create a multipoint ground. (b) Hybrid connection to ground. At DC or low frequency the inductors have no resistance and allow current to flow, creating a multipoint ground. At high frequency the raised impedance of the inductors blocks current and creates a single-point ground (after Ott, 1988)

the return current from relay-switching circuits. Racks are a source of noise, emphasized by their joints or slots and by pull-out drawers. Modern electronic circuits should not be grounded through racks but through their own ground connections. But if a rack or chassis has to be used as a ground connection, any seams, joints or openings should be inspected. Reliable electrical continuity should be provided across them by any feasible means, such as bonding.

Grounding of cable shields

Cable shields on cables used for low-frequency signals should be grounded at a single point when the signal circuit has a single-point

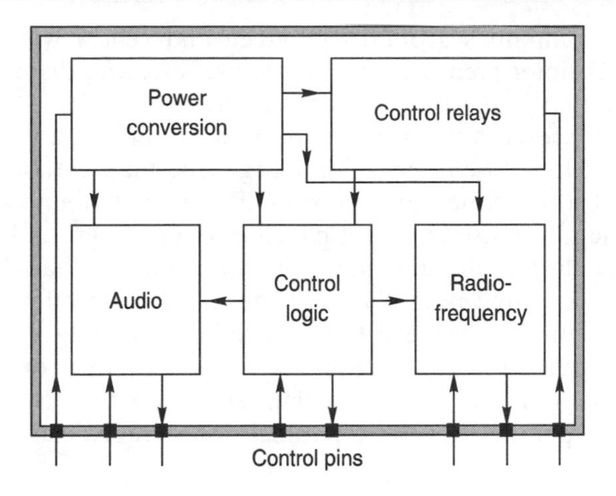

Fig. 12.4 Printed-circuit board, partitioned into sections so as to reduce interference between sections (after Mills, 1993)

ground. If the cable is grounded at more than one point, interference currents will flow, even along a coaxial cable or along a shielded, twisted pair of conductors.

Computer-aided designs for PCBs usually disregard EMC. EMC precautions must be added manually, e.g. by inserting a return conductor for a clock track (lead). Interference usually increases with rising clock frequency and with shorter rise times for clock pulses. Clock circuits and leads should be away from the input–output (I–O) area of the PCB. The I–O area should be a special part of the PCB and should include a 'quiet' I–O ground.

Thick cables have many different types of both shield and outer insulation. Underground mining cables for power may be double-wire armoured with two layers of spirally wound galvanized wire round them or single-wire armoured for cables with less severe duty.

No grounds, only a system reference

Electronic appliances often are connected not to ground but to a system reference instead. The appliance is then described as 'floating'. The system reference may be the metal of the frame or box or casing of the appliance but in any case all circuits in the unit are connected to it. This has the consequence that any loops formed by the wiring to it must be as small as possible. If they are not small they will attract interference in proportion to their size.

One very good reason for not connecting a signal circuit or a radio to the power company's ground (the green and yellow wire) is that it can radiate RF interference round the house, creating disturbance not only in one's own home but also next door.

A magnetic field crossing any ring of wire will induce a noise current in the wire which will be larger with a large ring than with a small one. Particularly at low frequencies one should aim to eliminate such loops which are generally caused by multiple connections to ground. A further reason for avoiding multiple connections to ground is that the earth is not always at the same potential, and the many connections therefore result in stray currents between the connections, creating noise.

Where a signal wire must cross over another wire or cable, the intersection should be perpendicular since if the two different conductors are parallel or nearly parallel, interference can result.

Miscellaneous sources of RFI (noise or interference)

Tracing a noise source with a portable radio always begins by listening to it at the highest frequency at which it can be heard. Higher frequencies are heard over longer distances than lower ones.

Corona noise usually at 30–50 kHz comes from leakages at the high voltages of overhead power lines or from similar high voltages normally found in TV sets, but it is noticed when the TV set suffers a fault, such as corrosion.

Underground transformers or transformers mounted on concrete pads at the ground surface sometimes produce RFI which is difficult to locate, at a frequency around 9 MHz.

A metal rubbish chute in a block of flats can act as an extremely disturbing source of RFI if its fixings are loose. The noise is usually silenced if all the fixings are tightened.

Some, fortunately rare, incandescent light bulbs can produce RFI. Very early bulbs that went out of production around 1925 were made of clear glass with straight wire filaments and a glass point on the bulb end. If they are still used in cellars or attics, they produce interference at 60–70 MHz.

Bonding

So far as EMC is concerned, bonding is the joining of metals so that the bond conducts electricity. It is best done by welding, brazing, sweating or soldering, although soldered joints are less strong than the others. Where these excellent electrical joints are impossible, bond straps must be relied on. 'Swaging' is joining by striking, which is popular because it avoids the damaging effects of heat.

If metal surfaces are machined, a satisfactory semi-permanent joint can be made. But machining is extremely costly, especially for large areas. Crimping makes good bonds if the paint and other coatings are removed before the crimping begins.

Threaded joints are not always reliable since AC tends to flow along the inductive circular route of the screw threads because of the skin effect and self-tapping screws especially are not to be relied on for avoiding interference.

The aluminothermic joint, often used for welding railway rails continuously, is very strong and reliable for steel and iron. A mixture of powdered aluminium with oxidizing agent and other substances is ignited to achieve a temperature of 2200°C which melts steel or iron. Aluminium also can be welded but the weld is not suitable for outdoor use.

Aluminium surfaces need a protective finish to prevent oxidation and pitting, but anodizing is not a conducting surface. Chapter 9, Shields, mentions paints which are electrically conducting.

Bond straps

So far as impedance is concerned, with DC only the ohmic-resistance of a bond strap matters, but as frequency rises the skin effect increases the impedance as well as giving the bond strap some self inductance, depending on its shape. Stray capacitance also arises between the bonded surfaces. Some US military specifications require a DC bond resistance of less than 0.1 ohm, occasionally less than 0.0025 ohm to minimize the dangers of electric shock. Radiofrequency affects the impedance of a bond strap at 1 MHz, equal to a large multiple of its DC resistance.

When electrochemically dissimilar metals are bonded together, the relative areas of anode and cathode are important. A large cathode, the source of electrons, involves more corrosion of the anode than a small cathode. Reducing the cathode area thus reduces the electron flow and the corrosion. Metals higher in the electrochemical series are more anodic and more easily corroded than metals below them. (See Table 12.1.)

Ideally the anode which corrodes should be replaceable. But in any case the anode should be protectively painted after the joint is made. The paint should cover the visible part of the joint between the dissimilar metals and as much of the cathode as possible.

Good bonding requires intimate contact between metal surfaces. The fastening should exert enough force to hold the contact in spite of shock, vibration, bending or tension. If dissimilar metals must be used, a replaceable steel washer between them is helpful, as is a protective paint after making the bond. Solder should not be relied on for

Table 12.1 Shortened electrochemical table (or galvanic series)

Most reactive	Least reactive
magnesium	graphite
aluminium	titanium
carbon steel	
iron	
lead	
tin	
tin–lead solders	
nickel	
chromium	
stainless steel	
copper	
noble metals, silver, gold, platinum	
cobalt	

strength. Bonds should be kept dry. A short discussion of bonding in cars to avoid interference to car radios is given in Chapter 8 (Impedance matching, balancing, tuning, interference, car radios).

Ideally good grounding

The most electrically-conductive and therefore the best earths (grounds) are the salt water of the ocean or of a salt marsh. Damp, mineral-bearing earth can be a good ground. The worst ground is dry, sandy or rocky soil without mineral content, which has very high electrical resistance.

Soil with 10% moisture has about 35 times the electrical resistance of soil with 20% moisture, and 70-times that of soil with 35% moisture which is effectively mud or marsh. Cold soil and, even worse, ice, is useless because of its high electrical resistance.

Ground rods ideally should not have a resistance higher than 25 ohms. If a rod has too high a resistance, several ground rods may be used in parallel, connected together at the surface with stout copper wire. They can be sunk about 2 m (6 ft) apart and should be 8 ft (2.5 m) long because the highest electrical resistance is usually found in the 6 ft (2 m) next to the surface, usually the driest soil. The labour and expense of sinking ground rods may be avoided if local bylaws allow the use of the metal cold-water supply pipe. This can be an excellent earth, provided of course that it is metal and not plastic. In or near cities it is essential to look into the local electricity regulations and to follow them carefully. Otherwise one may be prosecuted for making and using an illegal, dangerous earthing system.

Chapter 13
Health and other aspects of electromagnetic waves

Electrosmog

No one can doubt that electromagnetic radiation affects both the body and the psyche. It is used in so many ways to heal bones, flesh wounds, etc. (R. Santini, 1995). Helpful Schumann radiations, echoes of tropical thunderstorms, are reflected round the world between the ionosphere and Earth at around 8 Hz. A Schumann generator is installed for the ease of astronauts in their space vehicles. Whether electromagnetic fields cause cancer seems doubtful but many people have no doubt that once cancer has started they promote the disease.

Most conventional doctors do not acknowledge the existence of complaints known as electrical sensitivity (electromagnetic hypersensitivity) in spite of the many international conferences that have discussed them, including two in Copenhagen in 1994 and 1995 and a series on Work with Display Units that culminated in a conference in Milan, Italy, in 1994.

What is generally implied by electrosmog is pollution by electromagnetic radiation at the frequency of power supplies, 50 Hz in Europe and 60 Hz in the Americas. This is not the only possibility. Microwaves at frequencies of a million or more hertz also seriously harm people. Paul Brodeur explained this as early as 1977 in his book *The Zapping of America*, entirely devoted to harm from microwaves. Manfred Fritsch in his book *Gefahrenherd Mikrowellen* (Microwave danger, 1994) explains the risk in Germany. Radio waves at lower frequencies than microwave can also be harmful but every European home or business now has its own cable supplying electricity at 50 Hz, so 50 Hz is the most common European electrosmog and 60 Hz is commonest in the Americas.

West German experience

Fritsch felt deeply for the sufferings of people who lived close to overhead power lines. He spent two years, beginning in 1987, distributing questionnaires to people living within 150 m of overhead lines. With voluntary helpers he obtained more than 1000 forms filled in. Simultaneously he found out the facts from people living near railways. In west Germany they work on electric power at 16 500 volts and 16.7 Hz. Railway leakage currents passing through the ground near houses are believed to affect the health of residents nearby. Of his 1000 contributors only 3% had no complaints about health. There were often multiple complaints about these sources of radiation.

After only a year, one young couple began to suffer severely, the mother being the most distressed. But whenever she left the house for only half an hour to buy shopping she felt immediate relief. Her permanent headache left her, so did her general tenseness but they reappeared as soon as she came home. The family felt so strongly about the 20 000 V overhead line that they paid for a length of it near their house to be buried, involving many years of argument with the electricity distributor and costing them about 100 000 DM (£40 000).

Although the electric field is greatly reduced by burial, there is no such guarantee about the magnetic field. Magnetic fields penetrate earth and masonry so the buried magnetic field could be even worse than when it was suspended, especially if it is nearer. If the power phases are balanced, as they should be, as well as close together, the phases may balance each other out, resulting in an improvement, but much depends on cooperation from the electricity distributor.

Another German writer, Wulf-Dietrich Rose, technical director of IGEF (Internationale Gesellschaft für Elektrosmog-Forschung), an international association for research into electromagnetic pollution, gives in his book *Ich Stehe unter Strom* (1996) many examples of 'Elektrosmog' and how he overcame it. His list of ailments caused by this pollution includes headaches, lack of concentration, weakness, incompetence, fatigue, dizziness, impotence, rheumatism, various complaints of the heart or blood and changes of reaction time. Sleep troubles include disturbed or superficial sleep and the ability to sleep only when completely exhausted.

Buying all the latest electrical gadgets and installing many power sockets to supply them is likely to stimulate such afflictions because of their strong electromagnetic radiation (seeFig. 13.2). It is probably therefore not surprising that in German-speaking countries many people have suffered severely. They are technically advanced and prosperous. Until the late 1980s few people thought that electric installations could produce harmful radiation. But recently many paperbacks on Elektrosmog have appeared in German.

Railway leakage currents

W.-D. Rose, in the same book, relates two incidents in which strong leakage currents forced people to leave their houses although the magnetic fields were known to originate from the railway power system because they were at 16.7 Hz. In Frankfurt-on-Main a heart specialist with many electronic instruments moved from a remote suburb to his new surgery in the city centre only 600 m from the main railway station. In the new surgery he began to suffer from heart trouble himself and his instruments did not work so well as in his old surgery further out. His patients' ECGs (electrocardiograms) were different, their sufferings were more severe, and his computer often 'crashed'. The instruments and computer were taken away for testing and found to be in good condition.

Through the federal organization for radiation protection, the heart specialist discovered an electrical biologist who examined his new surgery and found it saturated with magnetic fields at 16.7 Hz. They exceeded 2400 nT (nanotesla) when a train was starting up, therefore demanding a high current, and this happened several hundred times a day. According to W. Maes' book (1995) and its Table (14.2), summarized in Chapter 14, a tolerable field strength in a bedroom is at only a hundredth of this level 20 nT (0.2 mG).

The doctor therefore cheerfully moved back to his old surgery in a suburb where his heart troubles and those of his patients diminished and his electronic devices worked well. He now routinely asks his patients how near they are to railways, overhead power lines, substations or other electrical equipment and often recommends them to measure the electromagnetic fields in their dwellings.

Munich Ostbahnhof

In Munich a house, nearly 1 km from the Ostbahnhof, owned by a large German bank was used by bank executives for short stays but at the time in question was intended for permanent occupation by one executive and his wife. In the previous seven years, four of those who had lived there had suffered heart attacks from which two of them had died and their wives had also suffered. The bank manager therefore thought it advisable to have the house examined. The normal field was found to be about 80 nT but when a train started up it rose to 450 nT. The manager therefore had no hesitation, he sold the house, knowing that such strong magnetic fields can cause heart troubles.

Many houses in Germany, as far as 2–3 km from an electric railway or tram line, suffer from such leakage currents which pass beneath them along underground watercourses or metal pipes or both. Sometimes such leakage currents enter the house along a metal pipe. This can be prevented by removing a short length of the metal pipe where it enters the house and substituting a corresponding length of plastic pipe. Any point on the pipe used for grounding (earthing) electrical devices in the house will then have to be moved 'downstream' of the length of plastic pipe, or another earthing point must be found.

Electrical sensitivity

Electrical sensitivity is not one complaint but many. Sufferers often have a history of chemical poisoning, whether from crop spraying or from mercury fillings in their teeth, etc. It probably affects 1% of the people in industrialized countries but other estimates reach 25%. Visual display units or terminals (VDUs or VDTs) are an important cause of these complaints, so are fluorescent tubes, according to FEB, the Stockholm support group for electrically sensitive people.

Such people suffer when they use electrical equipment, sometimes so severely that they cannot even use an old-fashioned ordinary fixed telephone. Although this is unusual, many more people find mobile phones extremely unpleasant, especially the modern GSM type which uses microwaves, pulse-modulated at 217 Hz. Electrical sensitivity often includes the appearance of skin troubles, itching, red spots, fatigue, weakness, headaches and much else. These troubles diminish when sufferers avoid using electrical equipment.

One unhappy ability, fortunately rare, is hearing the pulsation of alternating currents, reported in *Electrical Sensitivity News*, Sept.–Oct. 1996, in an article by Clarence W. Wieske. To most people the pulsation is inaudible. The article originally appeared as a paper to the First National Biomedical Sciences Instrumentation Symposium at Los Angeles on July 14–17, 1963. In Sweden, computer-related electrical sensitivity was also reported early but not before 1976. At the end of this chapter, an instrument, the Neurophone, is mentioned, which, when fed with radio frequency waves and in contact with human skin, enables the body to act as a radio set by converting radio frequencies to audio frequencies that are heard.

Alleviation

Several common ways of reducing the sufferings of electrically-sensitive people are:

- Fixing a grounded filter in front of a VDU screen to lessen its radiation.
- Connecting fluorescent tubes to ground (earth).
- Allowing only people who are not electrically sensitive to work at a computer.

Since all electrical gadgets emit electric or magnetic fields, usually both, electrically-sensitive people are advised to avoid them so far as possible. A consequence of this advice is that the two halves of the electrical industry, the very powerful generating part and the equipment makers, unite to deny any harm to health from electromagnetic radiation, opting to interpret results in favour of profits rather than public safety. The unfortunate result is that the electrical industry will not invest in impartial research. Although impartial long-term research into the health effects of electromagnetic radiation is urgently needed it seems unlikely that it will be undertaken, certainly not by the electrical industry.

Cellular Phone Taskforce

Although the electrical industry regularly affirms that the results of scientific research into the effects of electromagnetic radiation on health are uncertain, this is strongly denied by a recent writer, a holistic health practitioner since 1981, Arthur Firstenberg. In *Microwaving Our Planet* (1996), he shows, with some 230 references to the research and medical literature, that since the 1960s well-conducted, independent research has established without doubt the different sorts of harm done by many sorts of electromagnetic radiation, especially microwaves.

Arthur Firstenberg is chairman of the Cellular Phone Taskforce, in Brooklyn, New York, a citizens' group that opposes the uncontrolled growth of the cellular phone industry. The group combats in particular the invasion of a new digital system, PCS (Personal Communication Services), which has a top frequency of 2.4 GHz and requires closely-spaced transmitters placed on lamp-posts, dwellings, church steeples or trees. Like many other electrically-sensitive people, Firstenberg had to flee from New York when PCS was activated there in November 1996 because the microwaves made him seriously ill.

Help from electromagnetic radiation

Biofeedback

Despite horrifying talk of electronic warfare and of other harm to people done by electromagnetic radiation, there is a much more positive side to it. Properly applied, electromagnetic radiation can heal wounds or broken bones, help students pass exams, accelerate learning powers and creativity, cure drug addicts, relieve stress, help fat people become thin, thin people become fatter and much else. The improvements achieved are often permanent. It began with biofeedback in the 1960s before microprocessors existed. Consequently the instruments used (and they were essential) were expensive, cumbersome and available only in hospitals and similar institutions. This was the start of the idea of bodymind or mindbody, disposing of the idea that mind and body are separate. Anyone who practises biofeedback soon learns that the mind controls the body although the body also often controls the mind.

Originally the essential instruments included EMGs (electromyographs which measure muscle tension), EEGs (electroencephalographs which record brain voltages) and skin resistance meters which measure the electrical conductivity of the palm of the hand as described by C. M. Cade and N. Coxhead (1979).

Biofeedback teaches people to relax and that it is possible to control their bodies. Such training can alleviate phobias and anxiety and is said even to increase people's intelligence (IQ). Its possibilities seemed to be doubled or tripled when in the late 1970s at Berkeley, University of California, the neuroanatomist Marian Diamond surprisingly discovered through her years of research on rats that the brain becomes more powerful when its owner has a happy, entertaining, demanding life, full of challenge. Boredom, on the other hand, is bad for the brain. The more you learn, the more you can learn because the brain grows as it learns, quite the opposite of what was previously thought.

With the availability of microprocessors, the instruments for biofeedback became less expensive and more generally available, developing into the mind machines that now (1997) cost from £100 to £5000 or more.

Michael Hutchison, in *Mega Brain Power* (1994), describes many mind machines. He has acquired most of them and tried them all since 1984 when his interest in the subject started with his passion for floating in a warm bath of Epsom salts of extremely high buoyancy, described in his *Book of Floating* (1984).

Before discussing mind machines we should mention two other American writers who are as enthusiastic about mindbody or bodymind as Michael Hutchison. Sheila Ostrander and Lynn Schroeder's

Cosmic Memory (1993) is of great interest. These authors also wrote *Superlearning* (1979).

The brain's two hemispheres need explaining as well as the various brain waves recorded by the EEG. At least four, possibly five main types of brain wave can occur on the same EEG chart. They are measurements of the voltages of clumps of neurons which are perceived by the EEG electrodes on the scalp, after coming through the skull, so they are extremely faint – millionths of a volt. But they are fairly reliable.

According to *Mega Brain Power* (1994) the right and the left hemispheres of the brain do not always work together but when they do cooperate the mind works at its best. The body or bodymind is often controlled alternately by each hemisphere for about 1.5–2 h after which it hands over to the other hemisphere. Possibly the handover times are those when the mind works best. Although they are strongly linked and able to cooperate, the two hemispheres have different specializations. The left hemisphere controls the right side of the body, the right hemisphere controls its left side.

In 99% of right-handed people the left hemisphere is best at understanding or using words, it keeps appointments easily, is good at watching the clock, looking after details, being punctual, logical and analytical but it is slower than the right hemisphere. Although at least 60% of left-handed or ambidexterous people also have left-hemisphere language, up to 30% have mainly right-hemisphere language, according to Elmer and Alyce Green in *Beyond Feedback* (1977). The book also explains that biofeedback is a learning process and is not addictive. As soon as people have learned how to control their bodies through the biofeedback machines they feel no more need for them.

The right hemisphere quickly accepts large-scale and new information, is good at music, other arts and creative organization but is less cheerful than the left hemisphere and can suffer depression. But people differ enormously. All normal people have strong connections between left and right brain, so these distinctions between the two are only rough guesses.

John McCrone, writing in the *New Scientist*, 3 Sept. 1994, describes the 'entrainment' or 'frequency following' achieved by mind machines that was first used by neuroscientists in the 1930s working with the EEG and later in biofeedback. 'LS' (light and sound) machines achieve this by music and flashing lights sometimes accompanied by electromagnetic radiation at chosen frequencies.

One frequency of special interest is that of theta waves from 4–8 Hz, the twilight between waking and sleeping, common in children. This borderline state can release conventions of thought and bring new ideas. It may be of interest that sleep learning is not now approved by psychologists.

Binaural hearing

Considering binaural mind machines, the word binaural means 'two-eared'. When one of two headphones sends to an ear a frequency different from that reaching the other ear, the head hears the difference, a 'beat frequency' or drumming at the frequency difference. With a frequency difference of 8 Hz the beat frequency is 8 Hz. Robert Monroe's binaural Hemi-Sync tapes, by this means, aim to synchronize the working of the two hemispheres, but they also help people in their normal work.

Monroe's book *Journeys Out of the Body* (1971) describes his own experiences. When he realized he was up near the ceiling of his bedroom, looking down at his sleeping body, he was at first frightened but this fear disappeared as the episodes recurred. They intrigued him so much that he founded the Monroe Institute of Applied Sciences (Route 1, Box 175, Fabar, Virginia 22938) not far from his home near the Blue Ridge in Virginia, to help others achieve these altered states.

John McCrone describes the voltages perceived by the EEG electrodes as the background roar of the brain, the electrical noise generated by the firing of clumps of neurons.

The five main groups of brain rhythms or waves are beta, alpha, theta, delta, mu, starting with the highest frequencies and working downwards. Not all writers list the mu waves but they are mentioned briefly in *Scientific American*, Oct. 1996, pp. 61–2.

Beta waves at 14–40 Hz, occasionally higher, correspond to a normal active adult waking state, including anxiety feelings at the higher frequencies.

Alpha waves from 8–13 Hz belong to an adult state that is enjoyable and relaxed, probably indicating synchronism of the left with the right brain. Albert Einstein's more or less permanent alpha state in which he solved difficult mathematical problems seems to confirm the synchronism.

Theta waves from 4–8 Hz often arise with closed eyes, during deep relaxation between waking and sleeping, in a drowsy, twilight state sometimes accompanied by mental pictures. Theta also allows statements to be accepted as true, bypassing the brain's critical defence mechanisms. Zen meditators in the East need 20 years' experience to achieve theta regularly. Elmer and Alyce Green of the Menninger Foundation found in serious theta research described in their *Beyond Biofeedback* (1977) that theta rhythms improve not only health and human relationships but also learning abilities and originality of thought. Other researchers have successfully cured alcoholics by teaching them first to relax to an alpha state then to theta.

Delta waves at frequencies below 4 Hz are usually but not always produced during sleep, and often accompanied by healing as the brain releases quantities of growth hormones.

Mu waves are concerned with bodily movements such as smiling, chewing, swallowing. They diminish with movement or the intention to move.

Cosmic Memory (1993) describes the Lida machines captured by the USA in Vietnam and used by Soviet interrogators who had to talk to unsuspecting prisoners of war. The concealed pulses of its feeble electromagnetic field induce trance, elicit compliance and speed recall from memory, it is claimed.

Human direct current

Various researchers believe the body also has DC (direct currents) in addition to the alternating currents of the brain rhythms with their accompanying electromagnetic radiation. The DC is believed to command healing, cognition and bodily controls. Consciousness involves a DC in the brain which is reversed or stopped by hypnosis or anaesthesia. Dr R. O. Becker in the USA meticulously confirmed the presence of DC in the animal body by his investigations into hypnosis as well as into the healing of broken bones and in the regeneration of new limbs after amputations in salamanders and frogs.

All currents and voltages in the body are extremely small. The largest are perhaps those that come from the heart and are shown on the ECG (electrocardiogram) as millivolts. The EEG and EMG values are at a thousandth of this level – microvolts.

Acupuncture

Each of the brain's one million million nerve cells (neurons) has some 5000 connections with neighbouring cells, so the brain is a much more powerful computer than could ever be made by people. Although most of the body is self-healing, it can go wrong. About 5000 years ago doctors in China discovered acupuncture, which is healing by the insertion of needles at special points on the body. Not only does acupuncture improve health but it reduces or wholly eliminates pain during surgical operations.

The chemical anaesthetics used in the West have serious disadvantages especially for long surgery, so Dr H. L. Wen, a Canadian-trained neurosurgeon, went to China in the late 1960s to study acupuncture in anaesthesia. When he settled in Hong Kong in 1972 he used the accepted modern Chinese method of electroacupuncture which is convenient because there is no need for the continuous labour of needle twirling. Much to his surprise and delight, his patients told him that the electro-acupuncture had cured them of addictions to drugs such as morphine, heroin or opium.

Dr Wen naturally told his medical colleagues at the large Hong Kong hospital where he worked. One of these was the chief surgeon, Dr Margaret Patterson, who already used acupuncture to cure her painful migraines. Hong Kong is an ideal place for testing any method of curing addictions because the city has so many addicts. By 1973 the hospital had cured 140 addicts with addiction-durations varying from three up to 58 years. The doctors were sure of an important medical discovery but the Hong Kong authorities were not happy. They tried to silence them and frowned on any publicity about the curing of drug addictions.

According to her book *Hooked?* (Faber, 1986) the Hong Kong authorities' attitude forced Dr Patterson unwillingly to leave Hong Kong and to continue her research and healing work as a consultant in Harley Street, London. In Britain also, she obtained no research support from the authorities but her consulting work was fruitful. In 1974 she cured of his addictions Eric Clapton, one of the world's greatest blues guitarists who had been unable to perform for some years because of his drug dependence. Luckily he was happy to talk to the world about his recovery.

Dr Patterson thus came into contact with a number of pop musicians who enthusiastically boosted her healing method – NET (neuroelectric therapy). With Eric Clapton she had used needles but she discovered that these were less useful than the milliamps of electricity. One of the languages of the brain is frequency and each addiction has to be treated at its own special level. Narcotics need treatment between 70 and 150 Hz. Lower frequencies are suitable for barbiturates, sedatives and tranquillizers. Cocaine and amphetamines are best dealt with at frequencies as high as 2000 Hz. French doctors then were using frequencies with modulations as high as 100 000 Hz. Dr Patterson used modulations up to 50 000 Hz.

Apart from being an unusually conscientious doctor and researcher, Dr Patterson is also extremely able. At 21 she was the youngest woman ever to qualify as a doctor at Aberdeen University. Only four years later she became a Fellow of the Royal College of Surgeons, Edinburgh, an honour rarely achieved before the age of thirty. Her latest neuroelectric stimulator is a 9-volt pocket-size transistor device called the MegaNET 803, available in the UK from Neuroelectric Therapy (NET) Ltd, 50 Darnley St, Glasgow G4 12Y. Two wires lead to two electrodes behind the ears that provide milliamps of pulsed AC at any chosen frequency between 4 and 2000 Hz, and at any pulse-repetition rate suitable. The current is around 1 mA and rarely above

5 mA. Her treatment, neuroelectric therapy (NET), is, she insists, not a cure but only a detoxification, a removal of poisons. For a drug addict it has to be accompanied by a psychological–spiritual encouragement programme to change him or her from the lost, aimless person who became a drug addict.

Attempts with other CES machines (cranial electric stimulators) to copy the MegaNET have been unsuccessful because of the many variables in its programs – pulsing, frequency, etc.

The Brain Revolution, a mass of discoveries that began with biofeedback, affirms that mental states are linked with and determined by physical conditions in certain areas of the brain. For instance the right frontal cortex houses depressions but the left frontal cortex houses happy feelings. Many other accurately situated areas control other emotions.

Mind machines

Mega Brain Power (1994) describes many more brain machines than is possible in this chapter. One relatively simple type is the LS (light and sound) machine which resembles a Walkman combined with black sunglasses. The 'sunglasses' are in fact not transparent but contain two video screens, one for each eye. These are probably the commonest and cheapest, costing, in 1996, £100 or less. They have been claimed to cure allergies, anxiety, asthma, colour blindness, deafness, depression, epilepsy, fatigue, insomnia, migraine, pain, premenstrual syndrome, stroke and much else. When we think of the astonishing successes of the MegaNET in curing addictions which were not curable by other methods, we can happily suspend judgement. The Neurophone, another remarkable device, is mentioned later in this chapter.

Many other systems exist, including Ganzfeld, CES (cranial electric stimulation), light and colour devices and the expensive systems that involve movement of the whole body usually at not more than three revolutions per minute on a reclining chair or perhaps a sound (acoustic) table. They bring deep relaxation, encouraging entry into the alpha and theta states and enhance synchronization of the hemispheres.

Unidentified flying objects (UFOs)

According to Albert Budden (1995) the advent of UFOs is stimulated by electromagnetic waves of various origins, including those that come from earthquakes. A quite different view is held by Andrija Puharich, an American neurologist and writer who was a close friend of Uri Geller, the well-known Israeli spoon bender. Puharich's biography (*Uri*, 1973) of his friend includes fascinating accounts of his experiences with Geller in

Israel and elsewhere. He quotes no scientific theories but gives an interpretation of Uri's extraordinary powers. He affirms that Uri is under the control of the operators of these super-high-technology space vehicles. On several occasions when they were together, large, conspicuous UFOs in the sky were visible to them but not to anyone else. Many other miracles are confirmed by Geller's *My Story* (1975), which agrees with Puharich. It gives more detail in matters where Puharich must have felt that he could not risk his reputation as a scientist. Geller, not being a scientist, was much freer and his details are first hand.

Dangers

Ellen Sugarman (1992) describes the dangers of electromagnetic fields clearly. They are especially serious for children, being much more vulnerable than adults. She stresses that US school administrations, in the absence of authoritative information, are adopting the Swedish standard, that children should not suffer magnetic fields of more than 3 mG (300 nT). Her book includes 20 pages of summarized research into the human effects of electromagnetic fields, which is easily understandable to the lay person, as well as a glossary and useful appendices of facts about electrical devices.

Frequency-band windows

It must be admitted that radiation at power frequency (50 or 60 Hz) as well as at many radio frequencies is, or at least seems, harmless to people except those who are electrically sensitive. 'Seems' because nothing is known about the long-term effects of such radiation which is millions of times stronger than the natural radiation on the Earth. A window is a band of frequencies at which a specified effect can take place. Some animals, unaffected by low (or high) frequencies, may undergo an effect at a different higher (or lower) frequency, although a very slight movement of a kilohertz or less in either direction away from this frequency may completely annul the effect. Windows can also occur with radiation that is pulsed.

Pulsing

The pulsing used in radar searches is a concentration of transmission power (say) for an instant such as a tenth of a millisecond or microsecond every second or so, mainly to reduce power consumption. There is thus a pulse at a tenth of the transmission frequency, or any other repetition rate depending on the pulse spacing. Modification of radiation by pulsing can create a variety of windows, naturally, at frequencies always lower than the transmission frequency. It can even come close to an animal's heart rate, and this is only about 1 Hz. There are thus very many different frequencies in a particular signal. Perhaps the best known window (non-pulsed) on the vast electromagnetic spectrum is visible light, using the only wavelengths that we can discern with one of our senses, our eyes.

Ewald Kalteiss in his book *Elektrosmog* (1996) emphasizes that the future will bring yet more and more strongly-pulsed radio waves, but they may be harmless if their repetition rate does not approach that of the 'human' alpha, theta or delta waves or their harmonics. Alpha, delta and theta waves are from 1 Hz up to 30 Hz. Alpha waves are related to the Schumann waves at about 7.8 Hz which have already been mentioned.

EMERSETT

Sugarman (1992) reports a 1990 prosecution, a class action in favour of a group of people, which established two important points with serious implications for the whole of the electrical industry. First that electromagnetic radiation can be hazardous and secondly that workers must be told of any risks to which they may be exposed. This case was fought with the help of EMERSETT (Electromagnetic Radiation Case Evaluation Team), a cooperative on the US west coast, never afraid to confront the electrical industry's vast resources. Such litigation demands a high level of expert evidence, for which lawyers working alone do not have the information or funds. It is thus no longer possible for the electrical industry to say that it did not know about the effects of electromagnetic radiation.

The case, in Seattle, was on behalf of Robert C. Strom who suffered myeloid leukaemia because in the 20 years during which he had worked for Boeing in Seattle he tested missile components using pulsed radiation. Strom was awarded US$500 000 in an out-of-court settlement. With the money Strom set up a foundation to provide information about the dangers of electromagnetic radiation. Boeing also agreed to monitor the health of its workers in a medical program. The case enabled others employed by Boeing to claim for their sufferings from leukaemia and skin melanomas.

Sugarman points out also that litigation about electromagnetic fields is the fastest growing area of law in the USA, partly because of the authorities' refusal to set threshold limits that truly protect people's health. One consequence of the litigation is that values of property near overhead transmission lines in Texas lost 25% in value between 1990 and 1992, and much litigation now concerns property values. Houston Lighting and Power of Texas (HL&P) consequently, as a result of the court decisions, now warns its customers in writing about the electromagnetic risks of its overhead lines. In November 1987 HL&P had to listen in court to a statement about the possible cancerous effects of electromagnetic fields from overhead lines. The US NIOSH (National Institute for Occupational Safety and Health) is intensely interested in the safety of workers in the electrical and telecommunications industries, including TV, radio and amateur radio and the many unions in those industries share this interest.

Unions in the USA, the UK and Germany recommend wiring practices that reduce the spread of magnetic fields, including wiring in twisted pairs. Because electrical distribution in the USA often takes place on wooden poles that allow entry of the wires into an upper storey of the building, Americans usually use the term ‘transmission lines’ to distinguish long-distance overhead lines from the wooden pole-mounted distribution lines which are at a much lower voltage, and less high above the ground. These may nevertheless be as much of a risk for the local people because of the high currents they carry, releasing dangerous magnetic fields at peak hours, if as often happens they are close to houses. Sometimes the distribution poles carry a stepdown ‘pole-mounted’ transformer to bring the voltage to 110 V before the cable enters the house. The leakage from such transformers can introduce harmful, strong magnetic fields into a nearby house.

Sugarman believes that the solutions to the problems of electromagnetic fields in the USA will come slowly from persistent activity by consumer groups, a process that has already begun in Germany, and is perhaps most advanced in Sweden. Apart from the health of the public, she points out that energy conservation has to be observed and every extension of an overhead power line in the USA must be proved to be essential before it is allowed. Electricity distributors also have to encourage consumers to be energy efficient, and sometimes do so by providing financial incentives in addition to the money savings from smaller electricity bills. This ‘demand-side management’ (DSM) by the industry sometimes reduces power demand to such an extent that a projected new overhead line or power station becomes unnecessary.

Radiation at power frequency

Katalyse e.V. (1995) report experiences with infants up to one year old as well as with other small children. In bed they automatically try to get as far as possible away from any electrical equipment such as a radio alarm or a baby monitor. Katalyse interpret this as showing that children are sensitive to and dislike the feeling of man-made electromagnetic waves.

Since almost everyone’s home is connected to the electrical mains, the commonest radiation is that which comes from the mains at 50 Hz in Europe or 60 Hz in the USA. The wavelength at 50 Hz is 6000 km, and at 60 Hz is 5000 km. Consequently all activity concerned with mains power is in the near field. The distinction between near and far field is explained in Chapter 5, Antennas. In the near field the electrical and magnetic parts of the radiation must be considered quite separately, unlike higher frequencies such as RF, when the two act together. For example at 300 Mhz, a very high radio frequency, the wavelength is only 1 m, with most activity in the far field, at least for radio receivers. But since no transmissions are 'pure', harmonics exist at all wavelengths, and a 300 MHz transmission would contain harmonics at 600, 900, 1200 MHz and even higher frequencies.

Overhead power lines can carry enormous currents, thousands of amps, and they are ordinarily at high potentials, kilovolts (thousands of volts). The very high currents produce strong magnetic fields which are in one way more dangerous than the electric fields because they can penetrate much further, passing through brickwork and concrete without difficulty. It is possible to build a grounded shield that prevents the passage of an electric field, as explained in Chapter 9, provided that the grounding is effective. Trees between a house and a transmitter also will absorb any incoming electrical field so that it will not be felt inside the house. Any currents absorbed by the leaves or branches of the tree pass through the trunk to the roots and the damp soil around them. The sap in trees is an electrically-conducting liquid and every tree is effectively grounded through its roots.

With magnetic fields this is not possible, although some very expensive alloys can sometimes be used to shield against magnetic fields at lower frequencies. Mu-metal for example, made 15.5 cm wide as tape with a glued backing, although only 0.05 mm thick nevertheless costs per metre 180 DM (about £80) from a company called Db-electronic in Kiefersfelden, Bavaria. However its range is fairly limited since it can shield only at frequencies below 20 kHz, including any uniform magnetic fields generated by DC.

Alasdair Philips, editor/producer of *Powerwatch Network*, and a well-known EMC consultant, affirms that 0.8 mm is the thinnest feasible mu-metal but that it screens 50 Hz badly, with 1.6 mm being the smallest recommended thickness. Mu-metal is in any case too expensive except for the wrapping of very small units.

Electricity without payment

Overhead lines have many curiosities, some of them perhaps things which electricity companies would rather not have published, because they might enable people to use electricity without paying for it. The electric field under a power line is so strong that a fluorescent tube held below one by a grounded (earthed) person at night can be seen to glow. Another curiosity is wind action. When power lines sway, additional low frequencies are imposed on the basic 50 Hz frequency coming from the power station.

Human voltages

G. J. Wohlfeil (1995) describes his experiences as a building environmentalist investigating the pollution of dwellings by electric fields at power frequency. Figure 13.1 shows how he measures the voltage to earth (ground) of someone whose home is saturated by an electric field. The voltage is measured with an ordinary multimeter costing no more than £20. By this method he has measured AC voltages exceeding 20 V on people. He stresses that one's voltage to ground with bare feet should be zero. Even a heart pacemaker requires only 1 millivolt.

Wolfgang Maes (1995), confirming Wohlfeil, describes several experiences including one of an athletic man with an electrically-heated water bed that had several electrical faults. The man became seriously ill and was often taken to hospital by ambulance. In hospital he recovered but as soon as he returned home he became ill again. His wife complained about the electric shocks he gave her on the fingers or the lips when they came into contact. His voltage to ground measured 155 V and when his body made contact with an electric screwdriver, its bulb glowed brightly. After the electrical faults on his water bed had been corrected he recovered his health and his wife suffered no more electric shocks from him.

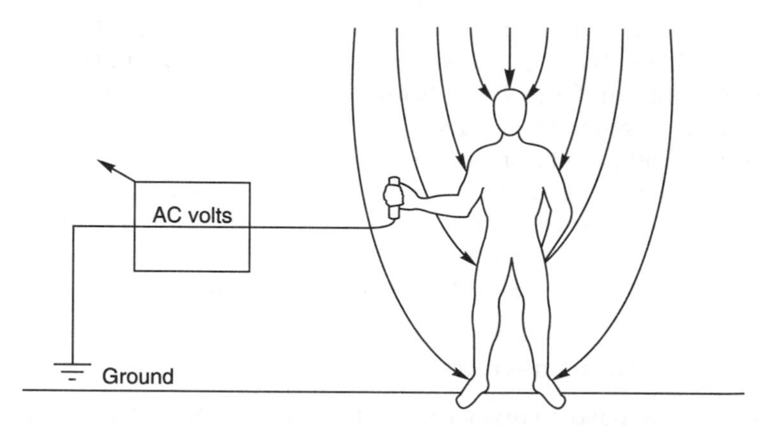

Fig. 13.1 Measurement of AC voltage induced in someone living in a dwelling polluted by an electric field (after Wohlfeil, 1995)

Maes mentions an even higher human voltage, moreover in the complete absence of any electrical fault. A sculptress who was the owner of a holiday villa on Lake Garda enjoyed the close proximity of her gas-fired heater which was next to her bed. However, every time she went to her villa she became ill to such an extent that she thought of selling the villa. This lady was measured as having a voltage to ground of 200 V, induced by the electric field emitted from the ignition device on the gas heater, because she enjoyed its warmth so much that she 'cuddled' it. After she realized that the plug of the igniter should be pulled out of the socket whenever the igniter was not in use, she lost the voltage and recovered her health. Figure 13.2 shows the strength of the electric field emitted by a power socket at various distances.

The 'human voltage' incidents related above concern the electric field at power frequency, which is effective even if no current flows. The magnetic field at power frequency is often present with the electric field but only if current is flowing. It has a quite different effect on the human body. It can penetrate the body as well as it penetrates masonry and is believed to create eddy currents in the conductors within the body. It is difficult to know what effect these have or where they are but they would seem to be well above the level of any currents normally present in the body. So they could be highly disturbing.

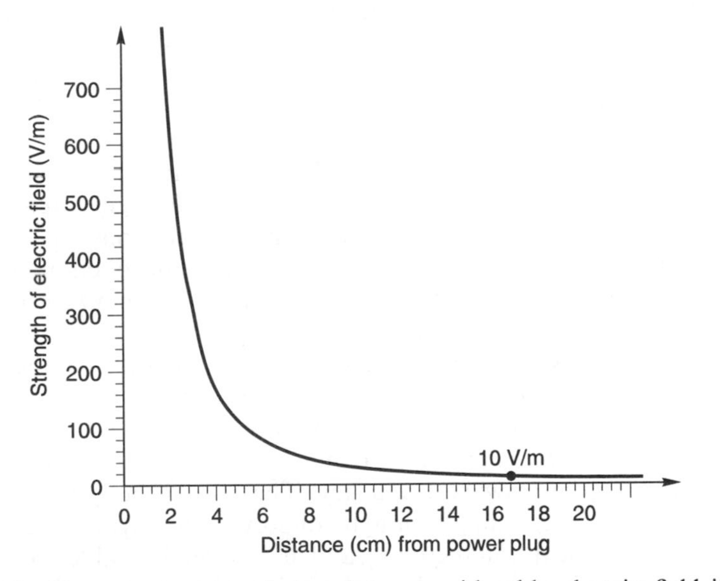

Fig. 13.2 Even a power socket emits a considerable electric field if you are close enough to it. The illustration shows how the field drops off with distance from the plug (after Katalyse, 1995)

German popular opposition

Germans, perhaps because of their country's outstanding prosperity during and after its post-war reconstruction, may have suffered more than others from electrical pollution (*Elektrosmog* in German). The German electrical generating and distribution companies operate in part under laws dating from Hitler's time when the dictator was preparing for war and aiming at *Autarkie* (self sufficiency). These laws allow them to decide the routes of their overhead power lines with little argument.

But Germany has many self-help groups for electrically-sensitive people. These tiny consumer groups heroically oppose the electrical industry, like David and Goliath, often successfully. Judges, at the instigation of such groups, have refused to allow the construction of radio towers and have forced a change of route for an overhead line. In contrast Britain has only two such groups. Germany's prosperity and high wages encouraged people at all income levels to buy every conceivable electrical gadget. No one wants to eliminate all electrical gadgets, they are too convenient. But we must be aware of their dangers.

Dangers of microwaves

Manfred Fritsch (1994) writes that in the present century German health authorities' statistics show alarming increases of many illnesses. In particular the incidence of heart trouble is 500 times what it was, with deaths from heart trouble 140 times what they were. Most of the increase has taken place since 1945 when construction of microwave transmitters became easier, especially after the great growth of microwave use during the 1960s. The microwave oven's frequency of 2.45 GHz thus became feasible and relatively economic because the cavity magnetron designed for it was in regular production. Users of microwave ovens should clean them so as to minimize leakage of dangerous microwaves through the joints.

Fritsch explains that the numerous tiny main units of the human body, its cells, are so small, nearly always well below 1 mm, that they can act as antennas to receive microwaves and even, very faintly, to transmit them. A wavelength of 1 mm corresponds to a frequency of 300 GHz, a top microwave frequency. The human body is some 70% water, usually containing enough dissolved impurities to give it reasonable electrical conductivity, and it is therefore able to act as a radio antenna which, naturally, can receive or transmit.

Concerning a more common frequency, that of power lines, the American writer Paul Brodeur, in *Great Power-line Cover-up* (1993), emphasizes the steep increases in brain cancer, lymphoma and melanoma in the USA and the rest of the world that have been caused by daily exposure to ordinary electricity. He tells the story of the EPA's (US Environmental Protection Agency) discovery that fields emitted by power lines promote cancer, and how it was prevented from publishing this discovery by the highest authority in the USA, the President's Office.

Fritsch stresses that the relationship between heart trouble and microwaves is confirmed by the statistics, which show that most of the illness occurs in the boom towns of Germany, i.e. the capital cities of the 17 German provinces (Länder) and the other prosperous cities saturated with microwave transmitters. The serious increase of German deaths from heart disease began between 1950 (14 000 deaths) and 1988 (140 000 deaths).

Several other German writers protest about the dangers of microwave cooking, that it spoils food and that the effect on human blood of eating microwaved food is to make people's blood resemble that of cancer patients (G. J. Wohlfeil, 1994; Heinz Steinig, 1994; Katalyse e.V., 1995). Deaths from cancer between 1950 and 1988 increased by 3.5 times, which, although large, is barely comparable with the increase in heart disease which multiplied tenfold in the period.

Building environmentalists

Germany's Green Party and its sympathizers have long made their country aware of every sort of environmental pollution. As a consequence, the profession of building biologist or environmentalist (*Baubiologe* in German, bau-biologist in the USA) arose for someone who is not only an architect or builder, but also a chemist and a medical specialist. At the end of the 1980s these well-informed environmentalists added electrical pollution to their list of specialities, and called it *Elektrosmog*. The German electrical industry, as would be expected, has issued pompous statements to belittle the concept of *Elektrosmog*.

At least two members of this profession have suffered severely from the harm done to them by overhead power lines. The books they have written are listed with other books at the end of the next chapter. Wolfgang Maes (1995), was forced to give up his job as a staff journalist because of ill-health caused by electromagnetism. However, after recovering, he studied *Baubiologie*, and now acts as a building environmentalist helping people who have trouble with their dwellings, as well as editing a periodical, *Wohnung und Gesundheit* (Dwellings and Health).

Heinz Steinig (1994) had to leave his profession producing water fittings because of ill-health, the result of overhead power lines near his house, and was lucky not to lose his sight completely. He taught himself *Baubiologie* and is now a full-time environmental consultant. His wife's health was also seriously impaired by the power lines. The heroic devotion of these and many other sufferers in avoiding distress to others is truly inspiring.

All electrical devices, sometimes even the very smallest, battery-operated ones, emit electromagnetic radiation, which can harm one's health. The first and easiest protection against them is by distance, e.g. 4 m from a TV set. A more expensive, laborious and highly technical option is to shield the device, enclosing it inside a protective, electrically-conducting surface that collects the radiation, as explained in Chapter 9, Shields. The shield must be earthed (grounded) to remove any electrical potential generated by the radiation that strikes it.

In several countries (USA, Canada, Italy, Sweden) consumer groups have been as successful as the Germans in demanding greater safety from the electrical companies. They have forced power lines to be re-routed away from schools or housing.

Swedish standards for computer screens

For some years now the Swedish standard for computer screens, known as the MPR2, has been accepted as the best available because it allows very little radiation to be emitted from a computer screen. In Germany it is recommended by the TÜV (official engineering consultant) of the Rhineland. But the MPR2 is being superseded by even more advanced Swedish standards. MPR is the Swedish abbreviation for National Council for Metrology and Testing, known since about 1987 as SWEDAC (Swedish Board for Technical Accreditation). For those who do not have either of these advanced screens, the radiation from a computer screen can be reduced by placing an earthed (grounded) filter in front of it which may be of metal-mesh or nylon so it only obscures the screen slightly. In Sweden the medical authorities insist that emissions from computers supplied to them conform to the strict standard MPR-90, summarized in Table 13.1 below.

Liquid crystal display (LCD) screens emit much less radiation than the usual cathode-ray tube, but even an LCD screen should be fitted with a grounded filter. A large screen radiates more than a small screen and a colour screen more than monochrome. An interesting article in the 1995 Copenhagen Conference on Electrical Sensitivity by Clas Tegenfeldt of Linköping University, Sweden, throws doubt on the 'perfect quality' of the Swedish MPR2 computer screen.

Table 13.1 MPR-90 standard

Frequency (Hz)	*Maximum allowable electric field*
5–2000	25 V/m at 50 cm distance in front
2000–400 000	2.5 V/m at 50 cm all round
	Maximum allowable magnetic field
5–2000	250 nT (2.5 mG) at 50 cm all round
2000–400 000	25 nT (0.25 mG) at 50 cm all round

According to Danze *et al.* (1995), EURELECTRIC (the European Council of Electrical Enterprises) has issued the following code of conduct for its members: 'If there is a reasonable supposition of environmental risk, efforts to reduce it should be made, even if the risk is scientifically uncertain.'

Benefits of magnetic treatments

An unexpected property of water is that it can be affected by a magnetic field, according to an American, Professor Kronenberg, reported by Wohlfeil (1995). Permanent magnets, fixed in pairs or fours round a pipe so as to attract each other with a north pole opposite a south pole, can reduce deposition of calcium or magnesium salts in the pipe and also help to remove deposits that exist already (Fig. 13.3). The 'magnetized' water also has better cleaning properties, and therefore saves washing powder. Wohlfeil reports, in addition, on American research that magnetic treatment of rocket fuel can increase the thrust of the rocket, saving 5% of petrol (gasoline) and 10% of diesel fuel. Magnetic treatment also improves the burning of natural gas, saving 12–17% in central heating burners or continuous gas heaters, and reducing the pollution from the exhaust. Other authors, for similar purposes, recommend a coil carrying DC round the pipe for magnetizing, instead of permanent magnets.

Mood control

Wolfgang Maes, a building environmentalist (*Baubiologe*), in his book *Stress Durch Strom und Strahlung* (1995) emphasizes that electromagnetic fields can affect people in the same way as alcohol or drugs, but without their being aware of the effect. Nerve impulses can be blocked or activated. The person who directs such control over people evidently can use the sufferer as a tool to do his bidding. Soviet psychiatric hospitals instead of drugs, often used one electromagnetic frequency to calm aggressive patients and a different frequency to

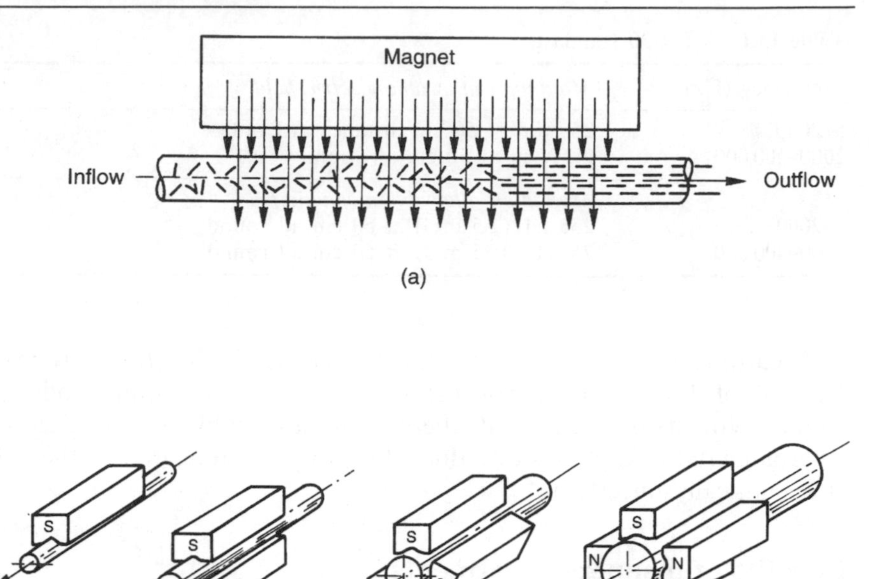

Fig. 13.3 Permanent magnets, set around feed pipes containing flowing water or fuel gas, reduce lime deposits in hot- or cold-water pipes, improve the effectiveness of water for washing and raise the combustion efficiency of fuel gases. (a) The magnets straighten the molecular or atomic orientation of water or gases and thus improve the washing ability of water and the combustion of efficiency of fuel gases. (b) The magnets, preferably more than one, should be arranged in pairs where possible to straighten the alignment of atoms and molecules (after Wohlfeil, 1995)

stimulate and encourage apathetic ones. Before the 1990 reunification of Germany, East German refugee specialists in such subjects were immediately snapped up by West German organizations. Western governmental research into such matters only began in the 1960s when it was realized that Soviet erudition was light-years ahead of the West.

Aromatherapy

Merely by filling their sale rooms with appropriate pleasant scents, luxury shops have for some years tried to induce wealthy people to buy more. This successful policy is beginning to spread to less lavish establishments. Aromatherapy has been used for centuries, although to heal and relax, rather than encourage sales.

As sales are essential it may not be long before we find that electromagnetism also will begin to be used to promote sales or simply to make people happy. Doctors have learned about many different human brain rhythms since Dr Berger of Jena described the first authentic EEGs (electroencephalographs) in the early 1930s. The alpha, beta, delta, theta and other brain rhythms dovetail with different states of human consciousness, including maturity or infancy.

If the patient merely closes or opens his eyes, raises or lowers a limb or modifies his breathing, the change is seen on his EEG chart and different brain rhythms accompany different moods. Little is known about mood control and what little is known has probably been kept secret but there are strong indications that mood control is practised by governments faced with protesters. The arrival on the printed page of such terms as 'non-lethal action', 'subliminal messages', 'synthetic telepathy' and 'cyborg' (cybernetic organism) shows that people know that something is happening, even if they do not know what.

The women who protested against the American nuclear weapons being held at Greenham Common airfield in Berkshire, UK, in the late 1980s felt certain that they were being subjected to electromagnetic radiation – that they were being 'zapped'. One reason for their certainty was based on what seemed to them the exact localization of the radiation.

An American consultant on electromagnetics boasted that on demand he could create any atmosphere in a group of people that was required of him:

> 'If you want people to stay in a bar and drink for a long time, I can do that. If you want people in a restaurant to eat quickly and vacate their seats for new customers, I can do that too.'

Very little if anything has been published on this fascinating subject, probably because research paid for by a government could not be released by the researchers. But an appropriate mix of a suitable carrier frequency with modulation can probably modify human moods.

Neurophone

The Neurophone is an invention of Dr Patrick Flanagan which, when in contact with the human skin, enables him or her to 'hear' music, speech, etc., transmitted to the skin at radio frequency. The neurophone has been through several versions, each patented by Flanagan in spite of obstruction by the US government which twice forbade him to work on it since they wished to keep it for the government's secret service. The first unit was invented when Flanagan was aged 14, in 1958. Some years later he applied for his first patent and was refused, and so, carrying a Neurophone and accompanied by his lawyer, he went to the US Patent Office to demonstrate it. Luckily the examiner had a deaf person available to test the device and when the deaf person, an opera lover, heard the voice of Maria Callas, tears streamed down his cheeks. The patent was allowed – the first time that a rejected patent application had been allowed.

Flanagan explains that human skin is a most important organ, enabling the body to act as a radio receiver, converting radio frequencies to something that is 'heard' in the brain. The radio frequencies perceived on the skin are well above the highest audio frequency but the body does the necessary down-conversion to audio frequency. For other 'mind machines' return to sections at the beginning of this chapter. Many different versions of the neurophone have been in production but the version proposed at the end of 1996 was expected to cost less than US$600, according to N. Begich (1996) in *Towards a New Alchemy*.

Support for the theory that the skin is the body's most important organ was published in the *New Scientist* of 7 Dec. 1996. *Atelopus zateki*, the golden frog, lives in the tropical forest in Panama, Central America. This frog, which has no ears but has been found to hear, was studied by Erik Lindquist and Thomas Hetherington of Ohio State University at Columbus, Ohio. They recorded on tape the calls of these frogs, played the tapes back to them and watched their responses. More than half of the frogs turned towards the sound from the tape and seven out of 15 answered back. Laboratory research showed that the frog's skin can vibrate at 2000 Hz, which is the frequency of their main call (*Journal of Herpetology*, **30**, p. 347).

Early research

We know very little even now about the way the human body works and reacts, and the early researchers can perhaps teach us something. Since 1892 when J. A. D'Arsonval, a French doctor and physicist brave enough to apply an electromagnetic field to himself, found that it warmed him, doctors and others have become intensely interested in such uses. D'Arsonval also noticed magnetophosphenes. In 1908, Nagelschmidt invented the word 'diathermy' (through heat) for the heating of the inside of the body by RF waves. This was 'long-wave diathermy' at 1–3 MHz, which was outlawed in 1954 by international agreement because it polluted the electromagnetic spectrum. By then it was almost superseded because after 1945 diathermy had already changed over to microwave frequencies, the very much shorter wavelengths of 915 and 2450 MHz.

Dr Harold Saxton Burr taught anatomy and neuro-anatomy in the Yale University School of Medicine for 43 years. His first scientific paper was published in 1916. His vacuum-tube voltmeter was the first that could measure the tiny millivolt DC potentials of living things without garbling the measurements, and was developed during his first three years of research. Over many years he observed the voltage differences between pairs of electrodes planted at 3 feet vertical spacing on tree trunks. But he was interested also in measuring human reactions, including those recalled under hypnosis. His book *Blueprint for Immortality* (1972) relates a person's emotion of intense grief that caused a large rise, of 14 millivolts, in 2.5 min.

An outstanding sequel to Burr's research came from a police specialist on lie detection in New York in 1966. Cleve Backster trained police officers to use sophisticated electronic equipment. In his evening research he had plants wired up to an instrument recording their electrical resistance and he established that a plant can respond to stress in a completely different species. In one room his equipment automatically dropped small crustaceans (shrimps) into boiling water. In another room his potted plants, connected to recording devices, showed a sharp response every time a shrimp was killed. When dead shrimps were dropped into the water, the plants gave no response.

Backster's plants responded equally strongly to an egg being broken in the room. In other experiments in 1972 he connected an egg to an encephalograph and dropped a second egg into boiling water 25 ft (8 m) away. Five seconds later there was a response so strong that the line almost jumped off the paper. These results, according to Dr Lyall Watson in his book *The Romeo Error* (1972), show that the most primitive living units, unfertilized cells, can communicate with each other, at least in emergency. Such communications might be electromagnetic but we do not know.

Medical engineering, ECG (EKG) and EEG

Out of the vast variety of electronic medical equipment based on electromagnetism, this book has space only for two units, the electrocardiogram (ECG or EKG), which is a graph of the heart beat, and the EEG (electroencephalogram), a graph or rather several graphs of rhythms in the brain. For the ECG, much of its information can be obtained more quickly and less expensively by someone who can take a pulse and use a stethoscope. The pulse rate, of 60–100 beats per minute (roughly 1 Hz), is not the only heart variable. Any of the important defects in heart rhythm can be felt on the pulse or heard in the stethoscope by a person with sufficient skill. The ECG, although costly, has the great advantage of being a written record which it is hard to dispute, so it is very widely used in hospitals. The ECG originated with Willem Einthoven (1860–1927) who coined the term 'electrocardiogram'. He was Professor of Physiology at Leiden University until his death and invented the 'string galvanometer', the first instrument to record the voltage changes of the heart muscle.

The EEG, however, is much more elaborate. Because its voltages must be discerned through the bone of the skull they are very small (microvolts) or one-thousandth of the millivolts that provide the ECG trace of the heart beat. A further complication is that the EEG normally uses 20 electrodes, set by the EEG technician in electrically-conducting fluid such as collagen at specified points on the scalp, to provide some ten different traces running more or less parallel.

Because of its tiny voltages the EEG must use the very latest techniques of satellite radio, designed to detect such voltages and in addition, to reduce disturbance to the EEG recordings, the patient has to lie down or sit without moving. To allow the necessary millionfold amplification for the EEG without intrusions, special precautions are taken to eliminate electronic noise. No fluorescent lamps are allowed in the EEG room and floor tiling and other furnishings are specially chosen to eliminate any possibility of static electricity. During an EEG session any electrical equipment unconnected with the EEG must not only be switched off, but their plugs must be removed from wall sockets. The patient's bed or chair as well as the EEG equipment should be grounded by a low-resistance lead to a good grounding plate and the resistance of the grounding lead should not be more than a few ohms. An ideal electrical ground is a metal plate 1 m square sunk into soil that is permanently damp.

The EEG recording takes at least 30–60 min, often much longer, especially if the patient is required to sleep during the recording. The graph shows the brain rhythms, which for a waking adult with the eyes closed are mainly alpha waves at 8–12 Hz. Many other frequencies are also revealed – the beta rhythm is 14–22 Hz, normal children have theta waves at 4–8 Hz and a person who is excited usually emits higher frequencies up to 30 Hz.

Epilepsy can be demonstrated to be present with certainty by an EEG. In the USA an EEG can be used to certify brain death if the waves disappear, making all ten or so lines flat for periods of ten seconds or more.

Each pair of electrodes indicates only the potentials discerned from the thousands of neurons (nerve cells) beneath it, but the fact that no evidence appears does not always indicate the absence of the condition sought. It may merely show that the neuron potentials are opposing each other. The EEG can detect abnormalities in the brain with reasonable certainty but gives little information about normal psychology.

Susceptibility of medical equipment

W. D. Kimmel and D. D. Gerke in *Electromagnetic Compatibility in Medical Equipment* (1995) explain some of the difficulties with electronic equipment in hospitals. Even if a shielded room is essential and it sometimes is, it is always expensive. The most notorious interference by high-energy medical equipment comes from the ESU (electrosurgical unit) which emits fierce broadband noise between 100 kHz and 100 MHz, typically from 500–1000 KHz. It is modulated at low frequency, hence its interference is broadband. Cathode-ray tubes in monitors of medical or other equipment suffer severely from ESUs because of their strong electric field strength of nearly 100 V/m which upsets most equipment.

Lasers also sometimes create real hardship. Some lasers have a strong pulse of 100 A with a rise time of 100 μs which is equivalent to a frequency of 3 KHz at millions of amps. Interference control of lasers is far from perfect.

Defibrillators have pulses involving a peak voltage of several thousand volts lasting less than 10 milliseconds and any electronic equipment connected to the patient must be able to tolerate such a transient. The same applies to the devices mentioned above and heart pacemakers are usually designed to accept such violence.

Telemetry is increasingly installed in hospitals and works at one of three bands around 160 MHz, 460 MHz, and 900 MHz. Another widespread source of interference to medical equipment is hand-held phones which are forbidden in some rooms but often allowed in corridors.

ECG, EMG, and EEG are difficult to protect by filtering because they all need bandpass at power frequency. They are not usually polluting because of their low power. The least sensitive, the ECG, is found everywhere in a hospital. It has 1 mV sensitivity at a bandwidth of 50 Hz. The others have microvolt sensitivities – the EMG has 100 μV at 3 kHz and the EEG has 50 μV sensitivity at 100 Hz.

Some examples of electrical pollution

When strong electromagnetic fields are brought near the body it might be expected that the body would be disturbed. One example is the electric sewing machine which emits such a strong magnetic field that a pregnant woman using one is believed to suffer a quintupled chance of her unborn child having leukaemia. In 1982 the German Berufsgenossenschaft der Feinmechanik und Elektrotechnik (Precision and Electrical Engineering Association) issued safety rules for protection against electromagnetic fields. They forbid the employment of pregnant women in any electromagnetic environment.

Harm to health from electromagnetic waves

Cot deaths

Britain is shocked by the unexplained deaths of some 2000 babies a year, with possibly twice as many dying in Germany. The subject was discussed in a parliamentary debate in March 1986. Most of the deaths occur in the winter months of high electricity consumption and largely among poor families who suffer from old wiring that is liable to release much radiation. R. Coghill's measurements at the cotside showed levels that were many times more than the normal electromagnetic field. Another name for these deaths is SIDS (sudden infant death syndrome).

In the USA as well as Germany many cot deaths have been noted near electric railways especially on the inside of a curve, where leakage currents in the ground from the railway can create high electromagnetic fields in nearby buildings. In several such instances, ground water near the railway intensified the field. A large metal water main became magnetized and added to the field. A nearby electric substation has been blamed because of the heavy magnetic leakage from its transformers.

Another possible cause of cot deaths according to W.-D. Rose (1996) may be the placing of a baby monitor too close to the baby. At 10 cm from the baby, the monitor's magnetic field next to the baby can reach 10 000 nT although at 30 cm this drops to 200 nT. He recommends a distance of 1 m or preferably 2 m.

Melancholy

In the 1980s, Dr Stephen Perry in Staffordshire noticed depressions which led to suicides among those of his patients who lived near buried power cables. Unaware of the presence of the cables, his patients suffered from the strong magnetic fields at the 50 Hz frequency of the heavy mains current.

At the very much higher frequencies of radio, many doctors were disturbed by the presence of a cluster of seven cancer patients, revealed in 1992, around the UK's most powerful civilian broadcasting station at Sutton Coldfield, near Birmingham. For TV and radio the total power input is almost 1800 kW without counting the long-wave and medium-wave transmitters.

Advice from a healer

A successful young man in a business in Essen moved to a new flat but found difficulty in sleeping as well as in getting his work done – troubles which he had never before experienced, so he was becoming desperate. After several unsuccessful visits to conventional doctors he spoke to a healer (*Heilpraktiker*) who advised him to have his flat surveyed for electromagnetic pollution. When this was done it showed a strong alternating magnetic field of 600 nT by the wall at the head of his bed. Behind this wall was a powerful pre-amplifier for the radio and TV receivers of all the flats in the block. Its unyielding magnetic field passed through the wall almost without reduction. It could be moved, although only one metre away, but a further reduction of its emissions was achieved by wrapping it with mu-metal foil. This nickel alloy foil, glued on the back, is extremely expensive by comparison with the aluminium cooking foil which shields against electric fields. But aluminium foil does not shield against magnetic fields. Mu-metal does, however, and can be equivalent in magnetic shielding ability to steel plate several millimetres thick. After treatment with mu-metal the magnetic field fell from the original 600 nT to 15 nT. The young man was then able to sleep well and resume his successful career.

Dental surgery's magnetic fields

Dental surgeons often suffer extremely high magnetic fields because of their electrical devices, such as the motor on the patient's chair, the dental drill and the halogen lamp illuminating the mouth. Some 5000–7000 nT is a common value for the magnetic field near the dentist's head, bringing a risk of heart trouble, high blood pressure, exhaustion, leukaemia, cancer and eye complaints.

The transformer for the motor controlling the angle of the patient's chair may leak a magnetic field of 2000–8000 nT which affects the legs and abdomen of the dentist. These strong magnetic fields may be reduced relatively inexpensively by moving the transformer away from the patient's chair. Extra but more expensive improvements can be obtained by wrapping the transformer or motor with mu-metal foil. Any halogen lamp should be supplied only by a twisted pair of wires to reduce the field emitted by the wires.

Brain tumours from mobile phones

Other devices which must be used close to the human body are mobile phones and cordless phones. These can harm the eyes and users have noticed after fifteen minutes of telephone conversation that they can see less well. Some of these devices that work at microwave frequencies are suspected of causing cataract and brain tumours as well as the eye troubles. The most powerful, up to 20 W, are those that are semi-permanently installed in cars. The federal German Bundesamt für Strahlenschutz (radiation protection) in 1992 issued a statement recommending minimum distances from the head at which the antennas of the different mobile phones should be held. The statement is reproduced in the accompanying Table 13.2, taken from *Katalyse e.V.* (1995).

Encephalograms made during a conversation on a GSM (digital) mobile phone have shown unusual peaks. GSM phones receive pulsed microwave at 0.9 GHz, with pulses 217 times per second. Cordless phones transceiving on lower frequencies are thought to be less dangerous.

The frequencies for GSM and DECT are the same throughout Europe. UK users are officially recommended to extend the antenna as far as possible so as to reduce the power demand of the phone. The French magazine *Santé* points out that these phones have a large power demand indoors, especially in a car, a lift or a concrete building.

Brain tumours from an aircraft cockpit

Airline pilots have a professional complaint, brain tumours, possibly because these men are surrounded by electrical instruments, largely at head level. In any aircraft everyone is confined in a metal case which reflects and thus may intensify radiation generated inside. Investigations at the Zürich Technische Hochschule (university) in 1990 confirmed the trend towards brain tumours.

Similar suspicions arise about the drivers of the fast electric trains in Germany who become extremely tired. Their troubles probably result from the very high currents in the electric motors of the locomotive and their resulting heavy magnetic fields. The Bavarian government Institute for Labour Protection and Health has therefore been

Table 13.2 Safe distances from the head for mobile phone antennas (german federal radiation protection department)

Frequency (MHz)	Maximum output (watts)	Minimum distance from head, approximately (cm)
450 – analogue	0.5	None (no danger)
	1	4
	5	20
	20	40
900 – digital (GSM)	2	None (no danger)
	4	3
	8	5
	20	8
1800 – digital (DECT)	1	None (no danger)
	2	3
	8	7
	20	12

measuring electromagnetic pollution at such work places since 1995. Fluorescent (strip) lighting and computer displays are also harmful. According to Dr R. O. Becker (1991) the fact that body currents are so tiny is a good reason to beware of any electronic devices that could inject fields or currents into the body.

Wulf-Dietrich Rose (1987) remarks that while weak electromagnetic fields can stimulate the body's activities, strong ones may inhibit them, adding that the difference between 'weak' and 'strong' may be only 200 μV. There are also big differences in the parts of the body that may be affected and in the many possible methods of applying electromagnetic fields. Generally, for healing purposes, weak fields are the most effective.

Healers

R. O. Becker is an American medical doctor with a considerable reputation from his research into healing. In his book *Cross Currents* (1991) he states that non-medically trained healers use electromagnetic radiation both for healing and for sensing the type and location of the complaint. He bases this theory on his friendship with a young Polish healer, Mietek Wirkus, who around 1986 came with his wife to the USA from Poland where he had been licenced and fully accepted as a genuine medical therapist, something which is forbidden in the USA.

Dr Becker stood erect fully clothed while Mietek Wirkus scanned his body by passing his hands over it a few inches away without contact but he detected some of Dr Becker's complaints. Dr Becker believes the transfer of energy during diagnosis can only be electromagnetic. Mietek Wirkus feels that energy passes from it to a patient during healing and that energy comes to him from the patient's troubled area while he is scanning it. Much more energy is required from him for healing a cancer or a schizophrenia than for arthritis, skin troubles, insomnia or neurosis. Although unfortunately no research has been done into the frequencies involved, Dr Becker thinks they are around 1 Hz and possibly up to 5 Hz.

Chapter 14
Health and measurements of electromagnetic fields

Right of way (ROW)

Inspired by a successful legal battle in 1983 in Houston, Texas, popular protests in the USA have forced powerful electrical companies to site their power lines away from schools and housing. It is normal in the USA for a strip of land, called the right of way (ROW), under overhead lines to be left without housing or schools. With higher voltages the ROW has to be wider. In New York State, no new power line of 23 000 V or more may be built within 50 ft of any dwelling. In California the following guidelines, quoted by Sugarman (1992), are used for the siting of schools near overhead power lines:

30 m from the ROW of a 100–110 000 V line,
46 m from the ROW of a 220–230 000 V line,
76 m from the ROW of a 345 000 V line,
107 m from the ROW of a 500 000 V line.

She thinks, however, that these distances are not enough and should be greater. G. J. Wohlfeil (1995) gives the following rule of thumb: for 380 000 V keep a distance of 380 m, and pro rata.

Wulf-Dietrich Rose in his book *Elektrostress* (1987) gives the following useful safe distances from overhead power lines for housing or schools used by people in good health. It can be seen that they are very much larger than those adopted in the USA, but these are the recommendations of his organization, the Internationale Gesellschaft für Elektrosmog-Forschung (IGEF), and are also much stricter than

Table. 14.1 Minimum distances between power lines and housing/schools

	Distance (m)	
Voltage (v)	(Germany) (W.-D. Rose)	(UK) (M. Payne, 1992)
380 000	180–250	225 for 400 kV
220 000	140–180	150 for 275 kV
100 000	80–120	90 for 32 kV
50 000	50–70	60 for 50 kV

German law requires. For people who have any heart trouble or who are sensitive to changes in the weather he recommends doubling the distances. In Table 14.1, M. Payne, in *Super Health* (1992), proposes the minimum distances listed on the right-hand side for the UK voltages shown.

Electricity without payment

Farmers normally use a 12-volt battery to provide an electric fence with the occasional jolts of high-voltage, low-current electricity. But if the farmer has an overhead power line crossing his land and if his electric fence is parallel to the overhead line he may not need the battery at least for the part that is parallel with the overhead line. According to Wulf-Dietrich Rose (*Elektrostress*, 1987) a 400 m fence wire could generate a current of 1 milliamp. He gives no voltage estimate because this must depend on the voltage of the overhead line and on its distance from the fence. Other things being equal, a 100 kV overhead line will provide only one-quarter of the voltage obtainable from a 400 kV line.

Farmers should bc aware of the danger of electrocution. A voltage induced in a fence near a high-voltage transmission line can reach 6000 V. It is possible to take power fraudulently from a nearby power line by placing below it an appropriate layout of coils and other electronics to lower the voltage thus induced to a level that is safe enough for house lighting. Electricity companies are generally not willing to provide public information about such matters. But for workers on overhead lines it is common practice to connect the line they are working on (which has been switched off) to a good ground (earth) so as to eliminate any danger to them from voltages induced by other lines which are still on power.

R. Santini (1995) points out that a French main road was jammed in January 1994 when an overhead power line broke and fell on it. Power

Maximum kilovolt	Maximum electric field below wires, (kV/m)	Distance from overhead line at which electric field drops to 1000 V/m (m)	Maximum magnetic field below wires, (μT)	Distance from overhead line at which magnetic field drops to 1 (μT) (m)	Minimum allowable height of wires at midspan (m)
380	5	60	15 at 1300 A	100	12
220	3.5	40	10 at 625 A	60	10
110	1.5	20	5 at 325 A	40	8

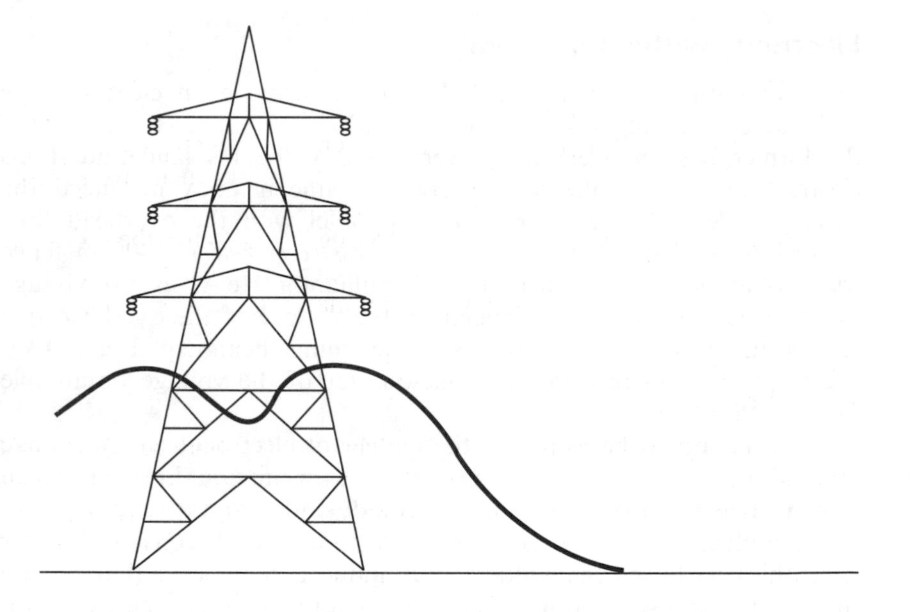

Fig. 14.1 Ground level electric and magnetic fields beside an overhead power line. The curves for the variation of electric field and magnetic field with distance from the centre of the overhead line are different. The curve shown is roughly that for the magnetic field. The figures in the table are rounded from Nimtz and Macker (1994)

lines directly over houses should be absolutely forbidden because of the danger of damage to the house and electrocution in the event of collapse.

Power lines over Fishpond village

In the hamlet of Fishpond, Dorset, in 1973, pylons of the National Grid were erected, carrying 275 000 V, and were later upgraded to 345 000 V. Lassitude and headaches were felt by most of the villagers after the erection of the overhead lines, especially on wet or windy days, but soon after the upgrading to 345 kV at least four people in the village suffered dizziness culminating in blackouts. A cycling visitor to Fishpond blacked out while riding his bicycle and broke some ribs. The presence of a strong electric field under these pylons can be experienced by anyone who holds up an unconnected fluorescent light tube below the power lines at night. It can be seen to glow. Of the 28 people living within 100 m of the overhead lines 21 suffered depressions, eye complaints, dizziness, or heart ailments, especially during rain or misty weather.

The story of Fishpond's electropollution is told in more detail by Hilary Bacon (1986) who lived there from 1973 to 1984. When the people of Fishpond complained, the electricity generating authority ridiculed them but the effect of this was to incite them to stronger protests. In 1984 Hilary Bacon was lucky enough to be able to leave Fishpond, and recovered her health. Strong support for the view that electromagnetic fields at power frequency can endanger health is given by Smith and Best (1989), as well as by many citizens' groups.

Soviet radiation

As a result of the Soviet 'woodpecker' broadcasts that began in 1976 and lasted for many years, people in western USA (Oregon) and Canada complained of high-pitched ringing in the ears, with pressure and pain in the head, anxiety, numbness, fatigue, insomnia and lack of coordination. They were heard all over the world except within the USSR. The extremely powerful Soviet woodpecker broadcasts went out at various frequencies between 1.60 and 17.54 MHz with ten pulses per second (10 Hz), sounding like a woodpecker. The estimated peak power of each pulse was 14 MW, making them some of the most powerful pulses of the time. The Association of North American Radio Clubs' Woodpecker Project coordinated listeners' protests from 18 countries at the February 1987 World Administrative Radio Conference for High-Frequency Broadcasting.

American secrecy

Unfortunately there has been a vast amount of secrecy on these subjects and not only on the Soviet side. The microwave irradiation of the US embassy in Moscow was known to the USA as early as 1953 and to every subsequent US government, but not made public until some 23 years later in 1976 when protective screening was installed in the building, reducing the field strength of the microwaves from 18 to about 2 $\mu W/cm^2$. The irradiation ceased in 1979 but reappeared briefly in 1983. Different writers quote other values for these figures.

Naturally no government discloses its reasons for secrecy but we can speculate on them. The USA, through its Central Intelligence Agency, the CIA, was, and probably still is, intensely interested in mind control by electromagnetism. Currently it uses pulsed microwave radar at 915 MHz in the USA in its 'mind control' activities. If it had disclosed what it knew about Soviet radiation of the US embassy, it might have revealed some of its own secrets, so instead it disclosed nothing. The purpose of the microwave irradiation of the US embassy is still unknown but Cyril Smith's comment in Ho, Popp & Warnke (*Bioelectrodynamics and Biocommunication*, 1994) suggests it may not have been malevolent but aimed merely to recharge the batteries of the Soviet listening devices installed in the embassy walls.

International Radiation Protection Association (IRPA)

Health in matters of RF energy is the concern of the IRPA, which states guidelines. Although usually not statutorily binding, these guidelines are very widely accepted.

RF waves are nonionizing radiation (NIR), unlike the ionizing radiations, X-rays or gamma rays, that result from nuclear explosions. Unlike NIR, ionizing radiation can create ions by adding electrons to, or tearing them off, an atom or molecule. The borderline between the two is in the ultraviolet region. The IRPA is concerned with both types of radiation and its INIRC (International Nonionizing Radiation Committee) looks after RF fields, providing guidelines at two levels, firstly at work (occupational) protection, and secondly the general protection of the whole population, including old people, pregnant women and children. The general protection is by far the stricter. Occupational health standards aim only to protect healthy adults aware of the risks and working under controlled conditions. Such people are often under medical supervision or at least monitored, so a lower degree of protection is allowed. A recent thorough literature survey by Arthur Firstenberg (1996) affirms that radio frequencies tend to damage the body in the same way as ionizing radiation.

DC power instead of AC

Because it is at constant voltage, with few variations of its electric current, DC generates no electromagnetic waves and so some of the health disadvantages of the AC supply can be avoided by using DC instead. DC can be generated from the AC supply without difficulty, with small losses but it is expensive to install because of the electrical accumulators and rectifiers needed and the space they occupy. G. J. Wohlfeil (1995) bluntly condemns any DC supply that is merely smoothed AC because it introduces more problems than it solves. He therefore recommends a DC supply based on 12- or 24-volt batteries or on solar cells. The batteries can easily be recharged from the 230 V mains. A battery supply has the advantage of being 'pure' DC but any DC supply involves the complication and heavy expense of two lots of wiring. He recommends AC for heavy loads such as the kitchen cooker, washing machine and fridge, keeping DC for all lighting circuits, especially the children's room.

An attractive halfway possibility, while also an expensive investment, is to use solar energy which is DC in the warmer months and to buy only energy-saving electrical appliances. According to Wulf-Dietrich Rose (1987) an average solar cell with an area of one square metre can yield 0.5 kWh per day from spring to autumn and costs some 2000 DM (£800). He thinks a one-family house with serious energy-saving ideas does not need more than six square metres of solar cells. But one must remember that he lives in Munich which is some 4° of latitude further south than London so has correspondingly stronger sun.

Alzheimer's disease, which in the USA is the fourth most frequent cause of death, can result from the physical stresses induced by the flickering light and other radiation from fluorescent tubes. In 1972 a Swiss doctor reported by Wulf-Dietrich Rose (1987) affirmed that this tube lighting also causes cancer by its emissions at 50 and 100 Hz and their harmonics.

Radar searching

Pulses sent out by radar systems are extremely powerful but also very short. They are followed by longer periods of zero release of power during which the electronics can receive an echo and from it interpret the distance, direction, and often the speed of any target found. The pulse can last between one and 10 μs and occupies only a small fraction of the cycle. A pulse may be heard as a click or a chirp or a knock and animals feel them fairly strongly. It is proposed that the maximum pulse power should be restricted even though the average power throughout the cycle is extremely low. A few people are able to hear

RF or microwave pulsing, but J. C. Lin in 1978 was probably the first scientist to report this ability. Lin wrote a chapter in R. Perez (ed.) (1995).

Human body temperatures

The frequency at which heating of the animal body begins is well below the 2.45 GHz used in microwave ovens. This frequency is often used by researchers, medical or other, because the essential hardware such as cavity magnetrons that generate the microwaves are already in production and can be obtained relatively cheaply. EM energy can heat the animal body but because of the differences between tissues (bones, nerves, blood, muscle, fat, etc.) it is hard to know where the heating takes place. Some membranes have extremely high resistance, but blood and other fluids are conductive. Heat itself is easily measured, but not inside the human body, where there is so much uncertainty. A further difficulty is that the quantity of heat transferred by EM energy is generally very small compared with that produced by the body itself. The human body's heat production averages 100 W, equivalent to that from a powerful incandescent electric light bulb. Healthy people are however able to produce much more; 3–6 W/kg body weight. (A small person weighs 50 kg.) In any case, all this heat is lost, 15–20% by breathing and evaporation from the skin, the remainder by radiation, convection and conduction.

The normal human blood temperature of about 37°C can safely rise to 39°C and often does, as a result of heavy physical exercise. Very few experiments on EM heating of the human body have been reported but much EM heating of animals has taken place at SAR levels above 1–2 W/kg. The data show that the development of an animal embryo is unlikely to be harmed by EM exposures that raise the mother's body temperature by less than 1°C. Combined with theory these results suggest that healthy humans when resting could suffer a whole-body SAR of some 1–4 W/kg for 30 min with a resulting body temperature increase of less than 1°C. It may be as well here to remind readers that the IRPA allows a maximum SAR at work of 0.4 W/kg.

RF sickness?

In the 1960s and 1970s Soviet and East European literature reported 'microwave sickness' among people exposed to pulsed radar. This sickness could not be confirmed but other complaints were corroborated. These were among people exposed to medium-wave RF (0.3–3 MHz) and among operators of machines for electrically welding and sealing plastics at a frequency of 27 MHz. The (mainly women) operators of these machines, some of them as powerful as 100 kW, were exposed to higher SAR than was recommended by the IRPA. The machines had to be redesigned with full shielding to protect the operators from RF waves.

Radio amateurs

Radio amateurs are people who after passing an exam to verify their understanding of radio have been granted a government licence to use a transmitter at low power and specified frequencies. They are represented in the UK by the RSGB (Radio Society of Great Britain) and in the USA by the ARRL (American Radio Relay League). An article in the RSGB's monthly journal *Radio Communication* in February 1982, 'RF hazards and the radio amateur', was written by two members of the UK National Radiological Protection Board, Messrs Blackwell and White. They point out that an average north European is nearly 2 m tall or half the wavelength of 4 m, which occurs at the frequency of 75 MHz. With his or her body correctly aligned to the source of the RF energy, in other words perpendicular to its line of approach, the body, which is an electrical conductor, may act as an antenna. It could then become resonant to this frequency and thus absorb ten times as much RF energy as it would in a different orientation. A small child or a rhesus monkey, only one third as tall as an adult human, would experience whole-body resonance at 225 MHz, three times the frequency for adult human resonance. With a vertically polarized wave the resonance would occur with a person standing erect.

Among the authors' safety recommendations for radio amateurs are:

- Never look down a waveguide.
- Replace and tighten any loose securing screws. Even a narrow gap under a cover can become an effective slot antenna if its length approaches a quarter-wavelength. Never peer through such a gap. If viewing or ventilation ports are essential, round holes are best. At 144 MHz, holes 2 cm diameter allow very little RF leakage.
- Never use a small antenna in the radio room as a dummy load. Instead use a properly screened dummy load. A small antenna can radiate considerable energy.
- Never adjust an energized antenna. Switch off first.

In connection with the first point above, a pilot who thoughtlessly stood in front of an airfield radar search transmitter for five minutes felt warming in the head and neck. Later other symptoms appeared, including loss of recent memory and extreme sleepiness. There are many reports of more serious harm from exposure to microwave pulses for as short as two minutes.

Damage to trees in Germany (Waldsterben)

E. E. Schulte-Ebbing, a forestry expert quoted by Wulf-Dietrich Rose in his book *Elektrostress* (1987), points out that in the two years 1982–4, forestry troubles in Germany rose disproportionately from the usual 8% to 50%. These two years involved a sudden growth of microwave transmitters, often connected with satellite communications. Until then atmospheric pollution by chemicals had been thought to be the cause of the tree disorders.

The fir trees, with needles 21–27 mm long, suffered most but spruces with needles about the same length also declined. The feathered leaves of ash trees suffered more than the heart-shaped leaves of lime trees. Similar troubles were noticed in North America even in pure mountain air and among the 90-metre-high giant softwoods. If we take the needles to be antennas 30 mm (0.03 m) long, this makes them likely to resonate to a frequency of 300 million m/s divided by 0.03 m which is a frequency of 10 000 million Hz or 10 GHz, a typical satellite frequency.

Schulte-Ebbing proved his hypothesis in one season, 1984, by placing wire mesh screens against electromagnetic waves round 15 of 22 distressed spruce trees near the Wendelstein transmitter which operates at 88–104 MHz. The 15 chosen spruces, averaging 3 m high, were each provided with four posts driven into the ground 50 cm away from the longest branches. Fine chicken wire with a mesh of 4 mm, fixed to the posts, was wrapped round the trees and in double thickness on the transmitter side, staying in place from early June to mid November 1984. Inside eight weeks 11 of the trees showed signs of distinct recovery with normal coloured needles and healthy crowns, although the other four showed no improvement.

The snag about this theory is that the wavelength of the Wendelstein transmitter at around 3 m for its 100 MHz does not in any way correspond with the length of the pine needles, at about 30 mm. Could it be explained by harmonics from the Wendelstein transmitter?

People watching birds flying have noticed that a large flock of many hundred birds can fly a wheeling turn with perfect agreement, however far apart. One theory is that the birds' feathers may act as radio antennas. Against this theory is the obvious fact that all birds have excellent eyesight or they could not land on twigs as they often do.

Safety against microwaves and RF energy

The WHO has published a number of texts on electromagnetic fields, some of which are mentioned below. In June 1996, however, it proposed an International EMF Project, to last about five years. Its purpose was eventually to publish scientifically-based recommendations for the health risks caused by exposure to electric and magnetic fields up to 300 GHz and including DC fields. Representatives of 23 countries and six international organizations took part in the EMF Project's preparatory meeting in May 1996.

Measures to ensure safety against RF energy especially microwaves, are listed by WHO EHC 16, *Radiofrequency and Microwaves*, 1981, with the following priorities:

1. Sound engineering design and construction.
2. Siting of transmitter.
3. Administrative measures such as warning signs.
4. Personnel protection, including restriction of entry to danger areas. In areas of strong magnetic fields, steel objects such as scissors, nail files, screwdrivers should be banned completely.

According to WHO, when the average power is small, not more than 0.1–1 mW/cm^2, workers exposed to either continuous or pulsed microwaves or RF energy are not in danger even with continuous full exposure over the whole working day. Many other people believe, however, that 1 mW/cm^2 is unsafe. But WHO EHC 69 (1987) ,*Magnetic Fields*, remarks that people working in an intense magnetic field of 2–5 mT near the hands and 0.3–0.5 mT near the chest and head suffer from irritability, fatigue, headache, heart complaints (increase or decrease in heart rate) and numbness. Unfortunately the many WHO reports (at least five) on EM radiation published between 1981 and 1993 largely ignore harmful effects at power frequency although they do have a long bibliography of medical research.

A WHO report by Suess and Benwell-Morrison (1989) points out that the sensation of pain on the skin begins to be felt at 45°C. However, at a much lower overall body temperature, 39.2°C, resulting from RF heating for 50 min, a healthy person would be sweating profusely and on the verge of collapse. (Normal temperature is 36–37.2°C.) A finger in contact with a charged object may burn when the current through it exceeds about 0.2 A.

Russian microwave safety thresholds

Probably because of their high investment in scientific research, the Soviet safety requirements against microwaves have always been one thousand times stricter than those in western Europe or the USA, with 10 μW/cm^2 set in 1970 against the western level of 10 mW/cm^2. The Russians also insisted on strict limitation of the time spent exposed to the radiation. Russian sufferers from long-term exposure underwent symptoms similar to those experienced at Fishpond. Similarly, the excellent German microwave ovens, although of very high quality, are unlikely ever to become a sales success in Russia. The leakage allowed from them in Russia is 0.01 $\mu W/cm^2$ against the German level, which is five times higher at 0.05 $\mu W/cm^2$. Quoted by Ellen Sugarman (1992), Dr R. O. Becker said that if the USA were to adopt the Russian radio frequency thresholds, nearly all its FM broadcasting stations would have to shut down.

Personal hazard monitors for NIR

The success of the hazard monitors worn outside the clothes of workers in or near ionizing radiation has prompted people in charge of NIR to devise a suitable monitor for their purposes. The need is for an alarm to sound when a pre-set power density is exceeded, especially for people checking newly-assembled microwave equipment including electronic countermeasures, whether in a detachable pod or in a ship or aircraft. Someone trained to use field-measuring equipment should be available to measure any field which raises an alarm. Such a hazard monitor should be extremely simple, small, lightweight, and easily attached to one's outer garment. It should not have measuring ability, but merely be able to raise an alarm and light an indicator lamp.

Several manufacturers, including Narda and Holaday, sell such NIR hazard monitors, according to Kitchen (1993). Inside one, on the surface next to the body of the wearer, is a layer of RF absorber to reduce the effect of EM energy reflected from the wearer's body. Different units cover the ranges 0.8–6 GHz (microwave), and 30–600 MHz (VHF and UHF) which might suit men who must climb antenna masts. The alarm can sound at either 10 or 50 W/m^2 corresponding to 1 or 5 mW/cm^2. Types for use in a noisy environment can be designed for latching, i.e. after the alarm sounds, an indicator lamp remains lit until the monitor is re-set.

Stationary hazard monitors for NIR

An alternative to the personal hazard meter worn by the operator is a NIR monitor in a fixed position in the work area. Such a unit can, without difficulty, be much more elaborate and heavier than the personal monitor. It could include in addition an infrared sensor to detect intruders at a time when the work area is normally empty and also switch on a warning sign or operate a door lock. It would always have an indicator of its battery voltage level and a visual alarm as well as the warning sound.

Narda make monitors with the trade name of Smarts. They sound off at a power density of 10 W/m^2 and work in the following frequency ranges: 2–30 MHz; 10–500 MHz; 0.5–18 GHz; and at 2.45 GHz.

Figure 13.1 shows one of the many ways of measuring the voltage of the human body. The NRPB (National Radiological Protection Board) in the UK objects to such methods which are obviously unsuitable for a moving person. The NRPB's device, clamped to the ankle, at least partly overcomes this objection and has performed well from 0.1–80 MHz and from 8–1000 mA.

Since the first astronauts returning from outer space suffered from the absence of the Schumann radiation that exists everywhere on Earth, NASA now installs artificial Schumann radiation on its spacecraft. Schumann radiation becomes much weaker at high altitude than at sea level, and this may perhaps explain jet lag.

Some bacteria, fishes and whales, possibly yeasts, as well as bees and all migrating birds, have a remarkable sensitivity to the earth's magnetic field which is probably not completely absent from humans. As far as the sensitivity of humans to strong magnetic fields is concerned, the available information is sparse. The daily (circadian) rhythms of humans have been found to be energized less by light than by the rhythms of Schumann radiation around 8 Hz. Scientists have found that birds can distinguish a frequency of 9 Hz from one of 10 Hz. The sure sense of direction of migrating birds has always astonished people but research has shown that birds use the dip angle of the earth's magnetic field, although it reverses as they cross the magnetic equator. They are probably also helped by the sun, while according to Dubrov (1974) the night-time migrators are helped by the stars.

Measurement of magnetic fields at power frequency

A meter used for measuring magnetic fields may be called a gauss meter, EMF meter (electromagnetic field meter), ELF meter (extra low-frequency meter), electromagnetic meter, magnetic scanner, magnetic flux meter, etc. Field strengths at power frequency (50 Hz in Europe, 60 Hz in the Americas) are the most important because in an industrial environment they are everywhere, and these are what we are now talking about. In some American publications the common unit for measuring magnetic fields is the older unit, the gauss (G) or milligauss (mG). In Europe the tesla (T), millitesla (mT), microtesla (μT) or nanotesla (nT) are more common. Therefore we first need to know how to convert tesla to gauss and the reverse.

Conversion factors

	1 tesla	= 10 000 gauss
or	10 gauss	= 1 mT (millitesla).
	1 gauss	= 0.1 mT = 100 μT (microtesla)
so	10 mG (milligauss)	= 1 microtesla (μT) = 1000 nT (nanotesla)
and	2 mG	= 0.2 μT or 200 nT (nanotesla)
also	1 tesla (T)	= 1 000 000 μT (microtesla)
	1 μT	= 0.8 amp/m
	1 amp/m	= 12.6 mG = 1257 nT

'Safe' threshold for power frequency magnetic fields

The value of 200 nT (2 mG) for magnetic fields at power frequency is important because it is often unofficially considered in the USA as the level that distinguishes 'safe' from 'unsafe' magnetic fields at power frequency. Children exposed to magnetic fields stronger than 200 nanotesla are thought to be twice as likely to die from cancer as those exposed to less than 200 nT .

This level of 200 nT (2 mG) has been semi-officially accepted in Sweden as a maximum allowable magnetic field for dwellings or schools, and in 1996 was beginning to be accepted in the USA but not in Britain. Many American authorities quoted by Ellen Sugarman (1992) prefer 100 nT while some even go as low as 50 nT. Sweden in 1996 reported a nationwide survey of domestic magnetic field levels. In cities they found 100 μT (1 mG) as a median value for homes and nurseries but in small towns and the countryside only half this level (50 μT or 0.5 mG).

Tolerable and intolerable field strengths

Table 14.2 shows the values of field strength in bedrooms accepted by Maes (1995), although it must be remembered that any value is liable to be distorted by the presence near the measuring instrument of people or metal objects. A good instrument will indicate frequency from, for example, 15 Hz to 100 kHz as well as the strength of the field. For magnetic fields a minimum of one hour's readings is needed in any room because of the large daily and instantaneous variations in current flow and consequently in the magnetic field.

The three different units that exist for a magnetic field are indicated in Table 14.2.

If there are either strongly fluctuating fields or radio frequencies or many harmonics, perhaps going up to kilohertz, the levels tabulated above should be divided by ten because higher frequencies may be less

Table 14.2 Electric and magnetic field strengths in bedrooms, tolerable and intolerable

Description (tolerable or not)	Electric field (V/m)	Magnetic field		
		nT	= in mG	= in A/m
Tolerable	1	20 nT	= 0.2 mG	= 0.016 A/m
Weak	1–5	20–100	= 0.2–1	= 0.016–0.08
Strong	5–50	100–500	= 1–5	= 0.08–0.4
Excessive	over 50	over 500	= over 5	= over 0.4

tolerable to humans than power frequencies. But the strength of broadcast radiation that reaches a radio or TV receiver is likely to be a small fraction of that from the power wiring and electrical devices in the house, unless the broadcasting transmitter is extremely close.

Power (flux) density

Below a frequency of 30 MHz, such fields are usually measured in volts per metre for electric fields or amps per metre for magnetic fields. Low-frequency EM waves have less energy than those of higher frequency, so a dividing line has been agreed at 30 MHz, above which the EM fields are estimated by power (flux) density, measured in watts per square metre (W/m^2) or milliwatts per square centimetre (mW/cm^2).

$$\text{Power (flux) density} = V/m \times A/m = VA/m^2 = W/m^2$$

$$1 \text{ watt} = 1000 \text{ mW}$$

and

$$1 \text{ m}^2 = 10\,000 \text{ cm}^2$$

so

$$10 \text{ W/m}^2 = 1 \text{ mW/cm}^2.$$

The relationships between power density (W/m^2) and electric or magnetic field strength are given by the following equations, assuming that measurements are made in the far field and in free space.

From Chapter 5, Antennas, Fig. 5.4, the wave impedance, V/A becomes 377 ohms (Ω) in the far field, in free space.

$$E = \text{electric field strength (V/m) and}$$

$$H = \text{magnetic field strength (A/m)}$$

$$\text{Power density, } P = EH$$

$$= \text{Electric field strength (E)} \times \text{magnetic field strength (H)}$$

and $EH = V/m \times A/m = VA/m^2 = W/m^2$.

Electric field: Power density, $P = E^2/377$, with the units $= (V^2/m^2)/\Omega = W/m^2$.

Similarly for magnetic field: Power density, $P = 377 \times H^2$, with the units $= A^2 \times \Omega/m^2 = VA/m^2 = W/m^2$.

Specific absorption rate (SAR)

The SAR (specific absorption rate) is a measure of the energy absorbed by a person or animal over a period of time, in watts per kg, NOT per cm^2 or per m^2. Because of its convenience, SAR uses the idea of a standard man, 1.75 m (5 ft 9 in.) tall, weighing 70 kg (154 lb) and with a total surface area of 1.85 m^2. If the total power flowing through a standard man is 7 W, the average whole-body SAR will be 7/70 = 0.1 W/kg.

It has been found that, neglecting cooling, an SAR of 2 W/kg for 30 min gives a temperature rise in the human body of 1°C, which is normally the most that people can tolerate.

The IRPA's recommendation, made in 1984 for whole-body exposure at work, was a maximum SAR of 0.4 W/kg of body weight. The general recommendation for the whole population was much stricter, one-fifth of the level at work, only 0.08 W/kg.

The narrowness of wrists and ankles intensifies the flows of any current through them, so effects there can reach 300 times the levels in the rest of the body. To limit the SAR in the ankles of a person exposed to an electric field at frequencies below 100 MHz, ANSI, the American National Standards Institute, together with IEEE in 1992 prescribed new limits. For frequencies between 100 kHz and 100 MHz the maximum total induced current for the feet is 200 milliamps for a controlled environment and 90 milliamps for an uncontrolled environment.

Metering and monitoring of electromagnetic fields

A monitor is a much cruder and less expensive device than a meter because it cannot measure. It provides only an indication of a particular level, usually one that is regarded as unsafe. In the nuclear industry it is worn on an outer garment.

Maes (1995) has pointed out that large errors (50% or more) can be caused in European 50 Hz measurements by the use of US meters designed for measuring 60 Hz fields. Radio frequencies, discussed by Kitchen (1993), are much more complex since they reach gigahertz levels and each instrument can be used only for its own selection of frequencies (frequency band).

Most instruments using isotropic antennas are more expensive but more reliable than non-isotropic (directional) ones. Any isotropic antenna has three parts, usually three arms at right angles to each other, therefore covering all three dimensions. The interconnected arms give a reading that averages the fields in all directions. A non-isotropic instrument gives a reading in only one direction, which can be unreliable. At the end of Chapter 2 is a short discussion of field-measuring instruments.

Earth's magnetic field (geomagnetic field)

The Earth's magnetic field is a constant which surrounds us from birth to death, and also to some extent controls us. According to Dubrov's fascinating book, *The Geomagnetic Field and Life* (1978), its variations affect the life of all animals. Although geological evidence shows from the magnetism of rock minerals, that the Earth has reversed its poles several times in the past, these reversals have taken place slowly over thousands or millions of years and for our purposes the geomagnetic field can be regarded as stationary, apart from its small daily, monthly and other variations.

In this it is quite different from the alternating and usually much stronger magnetic fields resulting from man-made electric currents at radio frequencies or power frequency. But superimposed on the constant geomagnetic field are even weaker alternations at frequencies from about 1–30 Hz, mainly around 8 Hz, known as Schumann vibrations. They originate from thunderstorms largely in the tropics and are spread over the world by echoing between the ionosphere and the Earth.

Apart from reversals of polarity, the magnetic north and south poles are in continual slow movement and now change by 7.5 minutes of arc every year. The north magnetic pole has been west of true north for three centuries but in the 1500s it was east of true north. In 1996 magnetic north at Birmingham (UK average) was 5° west of true north but in the year 2000 it will be 4.5° west. This rate of change is slowing down. Ordnance Survey maps publish the magnetic variation at the time of printing, obtained from the UK Geological Survey.

The strength of the Earth's magnetic field averages 50 μT (500 milligauss, mG), being 35 μT at the equator and 65 μT at the magnetic poles. The field is horizontal at the equator and vertical at the magnetic poles (dip angle 90°). Between these extremes the geomagnetic field can be in part described by its dip angle, which in northern Europe is around 50°.

Reduction of damaging fields

Damaging magnetic fields are also emitted by overhead projectors (3000 nT), electric sewing machines (2000–12 000 nT), electric under-floor heating (6000–20 000 nT) and many other devices. If switching off the power is impracticable, three methods of reducing the magnetic field exist – keeping one's distance, wrapping the device with mu-metal foil, and in the event of a wall or floor intervening, coating this either with magnetically protecting paint and/or with mu-metal foil.

Prevention of electromagnetic pollution

Sooner or later people will have to think deeply about how to reduce electrical smog. Some German families have even paid for an overhead line near their house to be put underground. This stops or greatly reduces the electrical field but does very little against the magnetic field unless the currents in the buried lines are more in balance than they were overhead. If the underground cable is nearer the house than the overhead line was, the magnetic field in the house may even become stronger. Any overall improvement must wait until the population as a whole is aware of the dangers of electrical smog because such changes will have to be drastic and therefore expensive.

Broadcasting transmitters have to remain, although their strength could probably be greatly reduced if broadcasting organizations in all countries were to agree to a general reduction of broadcasting power. This would yield appreciable energy savings, involving large annual economies.

So far as telephone and data links are concerned, these can be operated by optical fibres, which have the enormous advantage of high bandwidth. They also neither create nor suffer from electromagnetic interference. They are widely used in aircraft because of their small weight, a tiny fraction of the weight of copper. But the absence of interference is also important for aircraft. On the ground probably the main disadvantage is the cost of burying the line.

Energy transport without electrical pollution

An even longer term proposal which may eventually be realized is the disappearance of high-voltage overhead power lines and their replacement by high-pressure gas pipelines which of course create no electromagnetic pollution, although environmentalists unconcerned with energy needs might object to pollution caused by their construction and burial.

Long-distance gas transport at some 70 atmospheres pressure is efficient, with low losses and is practised throughout Britain by British Gas. Before the discovery of North Sea gas in the mid 1960s Britain imported liquefied natural gas from Algeria to Canvey Island on the Thames Estuary, and piped it to Leeds at high pressure, a technique which was later developed for moving North Sea gas. The pipes being buried, unlike overhead power lines, are not in danger from snow and frost. The original reason for building overhead power lines was that electricity was cheaper to transport than coal but the situation has changed. The cheapest way of transporting energy over land is probably now by high-pressure gas pipeline. By sea, bulk transport of liquefied petroleum gas at low temperature in ships is even more economical.

Precautions against electromagnetic pollution

All authors agree that distance is the first human protection against radiation although some writers insist on larger distances than others. Secondly they agree that any electrical gadgets as well as loose unscreened cables should be reduced to the absolute minimum, especially in bedrooms because people asleep are at their most vulnerable. The minimum distance from any of the above and from plumbing pipes or from cables hidden in walls is 1 or 2 m. Only screened cables should be used, and the screen should be grounded at one place. Copper braid screens on cables are flexible and obtainable on cables in the UK. Power cables should normally be 3-core, although 4-core are used for connecting to two-way switches but in either case one core is for grounding.

Electrically heated cushions or blankets should be absolutely forbidden for pregnant women and fluorescent tubes should be kept out of bedrooms, and preferably also from the rest of the house. Electric floor heating is highly undesirable, so are dimmers for lights, and ring main supplies. Radial connections from the power cable are preferable to ring mains because they greatly reduce both magnetic and electric fields. All plugs should be pulled out of their sockets whenever a device is not in use. Only first class electrical installers are worth employing, never do-it-yourselfers. A fully qualified building environmentalist should be consulted about any problems.

In a block of flats, people must protect themselves against their neighbours' electric gadgets that have a high current flow and consequently a heavy magnetic field that may pass through a wall. A protective grounded shield may be made of the very expensive sticky-backed mu-metal foil referred to in Chapter 13. Any shield should be grounded. If protection against only an electric field is needed, aluminium cooking foil or an electrically conductive paint or curtaining may be used, but this also must be grounded.

Any computer monitor bought should be to the Swedish MPR2 standard or better. The TCO is a recent Swedish improvement on the MPR2.

Devices with transformers should not be used. If they are unavoidable one should keep 1 m away from them or convert the device to battery operation. Car phones, smaller mobile phones, walkie talkies, cordless phones and other transceivers are undesirable. If a mobile phone must be used, the older type may be preferable, being analogue, at lower frequency and not pulsed like the modern GSM or DECT types, although their power output is often higher.

Any metal in the house's structure or insulation should be grounded. This applies even to plasterboard or other sheets with aluminium foil surfacing. Water pipes and heating pipes where possible should be avoided in bedrooms because if incorrectly installed they may carry stray currents causing disturbing electromagnetic fields.

Demand switch

Already well established in mainland Europe, a demand switch (*biorupteur* in French, *Netzfreischalter* in German) is one that, when the last light has been switched off, automatically disconnects the 230-volt AC supply from any chosen circuit, and substitutes a DC supply at 2 or 6 volts. As soon as the first electric light is switched on in the morning, the 230-volt AC is reconnected and the DC supply cut off. In Germany in 1996 such a switch cost 220 DM, and its installation by a competent electrician cost another 80 DM, totalling 300 DM or about £130. The DC supply has no current flow so emits no magnetic field, and its electric field is negligible. The demand switch does provide a peaceful night's rest in all bedrooms, but for those who cannot afford it, an equally effective, although laborious, solution is to pull out the fuses of the circuits to bedrooms every night and to reinsert them in the morning.

Warning: any circuit controlled by a demand switch should not include kitchen equipment such as a fridge or any other device which switches on and off frequently.

Manfred Fritsch (1994), a building environmentalist who has befriended many people dwelling near overhead power lines, specifies the following minimum distances from beds:

- 1.5 m from unscreened cables,
- 2 m from radio alarms, stereos, battery-chargers, low-voltage lamps, or from a fuse box, rising-main cable, house distribution panel, night storage heater, telephone, or baby monitor,
- 3 m from a black and white TV set,
- 4 m from a colour TV set.

He adds that any standby operation of a TV or stereo is undesirable and that children should not play with light switches because of the high transient currents and resulting heavily-polluting radiation during switching on or off. A battery-driven radio is better than a mains-driven one and a clockwork clock is better than a battery-driven one which, naturally, is better than mains drive.

Wolfgang Maes (1995) broadly agrees with Fritsch but goes into very much more detail, classifying five types of pollution and listing precautions against the following:

- Power-frequency electric fields,
- Power-frequency magnetic fields,
- Fields at radio frequencies,
- Electrostatic discharge (ESD),
- Distortion of the Earth's magnetic field.

Maes thinks that no dwelling should be nearer than 200 m from any overhead power line or electric railway. Electrical substations, because of the heavy magnetic field leakages from their transformers, are also most undesirable neighbours. Any window facing an overhead power line or transmitter for radio or TV should be screened with electrically, conductive curtaining which is grounded (earthed). Any site for a dwelling from which a radio or TV transmitter can be seen is undesirable according to Maes. He recommends brick, stone or concrete blocks as building materials for housing because they impede the electric field.

The house mains supply should include an electric filter that minimizes higher frequencies. Baby monitors that use radio should be avoided, at least in bedrooms. Any other baby monitor used should not be nearer the baby than 2 m. Distance is often the easiest way to minimize dangerous radiation, and for a magnetic field it may be the only way. An electric fan on a desk should not be nearer the body than 1 m. For electric typewriters and razors, distance is not possible. People, who are uncertain about whether these very strong magnetic fields may be harmful, should avoid them.

Maes has made many measurements in cars and in German trains. Cars with the most electrical gadgets have undesirably strong fields. Frequencies in a moving car that are hard to explain may be caused by rotating magnets such as wheels or drive shafts. These units can easily become magnetized and, turning at different speeds, will produce AC potentials at different frequencies.

Germany's fast, highly efficient intercity trains, always spotlessly clean apart from their invisible electromagnetic pollution, subject their passengers to intense AC magnetic fields at 16.7 Hz, varying typically from 1000–34 000 nT and averaging 21 000 nT, he has found.

Bibliography for Chapters 13 and 14

Books concerned with life

Backster, C. (1968). Evidence of primary perception in plant life. *Int. J. of Parapsychology*, **10**, 329.

Bacon, H. (1986). Hazards of high-voltage power lines. In *Green Britain or Industrial Wasteland?* (Goldsmith and Hildyard) Polity/ Blackwell.

Becker, R. O. (1991). *Cross Currents*. Bloomsbury Publishing.

Becker, R. O. and Marino, A. A. (1982). *Electromagnetism and Life*. State University of New York Press.

Becker, R. O. and Selden, G. (1985). *The Body Electric, Electromagnetism and the Foundation of Life*. Morrow, NY.

Begich, N. (1995). *Angels don't play this HAARP*. Earthpulse.

Begich, N. (1996). *Towards a new alchemy*. Earthpulse Press.

Bennett, Jr, W. R.(1992). *Health and low-frequency electromagnetic fields*. Yale UP.

Bone, S. and Zaba, B. (1992). *Bioelectronics*. Wiley.

Brodeur, P. (1977). *The Zapping of America; Microwaves, their Deadly Risk, and the Cover-up*. W. W. Norton & Co. Inc.

Brodeur, P. (1993). *The Great Power-line Cover-up. How the utilities and the government are trying to hide the cancer hazard posed by electromagnetic fields*. Little Brown.

Budden, A. (1995). *UFOs, the Electromagnetic indictment*. Blandford.

Burr, H. S. (1972). *Blueprint for immortality, the electric pattern of life*. Neville Spearman.

Cade, C. M. and Coxhead, N. (1979). *The Awakened Mind, Biofeedback*. Wildwood House.

Coghill, R. (1990). *Electropollution, how to protect yourself against it*. Thorsons.

Coghill, R. (1992). *Electrohealing*. Thorsons.

Danze, J. M., Le Ruz, P., Santini, R., Bousquet, M. and Mercier, J. L. (1995). *Pourquoi et comment mesurer les champs électriques et magnétiques 50/60 Hz*. Encre.

Dubrov, A. P. (1987). *The Geomagnetic Field and Life*. Plenum. From the Russian (1974) *Geomagnitnoe Pole i Zhizn*, with 1228 literature references.

Duchene, A. S., Lakey, J. R. A. and Repacholi, M. H. (eds) (1991). *IRPA Guidelines in Protection against nonionizing radiation*. Pergamon.

Endler, P. C. and Schulte, J. (eds) (1994). *Ultra high Dilution Physiology and Physics*. Kluwer Academic.

Evans, H. (1989). *Alternate states of consciousness, Unself, Otherself and Superself*. Aquarian Press.

Evans, H. (1993). *SLI Effect (Street lamp interference)*. ASSAP.

Firstenberg, A. (1996). *Microwaving our Planet, the environmental impact of the wireless revolution*. Cellular Phone Taskforce. (PO Box 100404 Vanderveer Station, Brooklyn, NY, 11210, USA)

Fritsch, M. (1994). *Gefahrenherd Mikrowellen. Infarktrisiko und Gesundheitsgefahr durch Sendeanlagen, Mobilfunk und Mikrowellenherde. Der Lebensbedrohende Elektrosmog*. Ehrenwirth. (Microwaves a source of danger. Threat of heart attacks and ill-health from transmitters, mobile phones, and microwave ovens. Electrical pollution menaces life.)

Fritsch, M. (1994). *Ein Leben unter Spannung, krank durch Elektrizität. Der alltäagliche Elektrostress, Schutz vor Elektrosmog*. Ehrenwirth. (Life under tension, illness and everyday stress from electricity. Protection against electrical pollution.)

Geller, U. (1975). *My Story*. Robson Books, UK.

Grant, L. (1995). *The Electrical Sensitivity Handbook*. Weldon Publishing. (PO Box 4146, Prescott, AZ 86302, USA)

Green, E. and Green, A. (1977). *Beyond biofeedback*. Delta, NY.

Ho, M.-W., Popp, F. A. and Warnke, L. (eds) (1994). *Bioelectrodynamics and Biocommunication.* World Scientific, Singapore.

Hutchison, M. (1994). *Mega Brain Power.* Hyperion.

Kalteiss, E. (1996). *Elektrosmog – im Zweifel für den Angeklagten. Was die Natur uns bietet und was sich der Mensch selbst zumutet.* Verlag Elke Hensch. (Electrical pollution – doubts for the accused. What Nature offers us and what humanity itself expects.)

Katalyse e.V. (1995). *Elektrosmog. Gesundheitsrisiken, Grenzwerte, Verbraucherschutz,* 3rd edn. C. F. Müller, Heidelberg. (Electrical pollution. Health risks, allowable limits, consumer protection.)

Kitchen, R. (1993). *RF Radiation Safety Handbook.* Butterworth-Heinemann.

Kreitz, H. (1996). *Krankheit und Tod durch Elektrosmog, Sonne und Wasseradern. Ursachen denen man ausweichen kann.* Herold Verlag Dr Wetzel. (Illness and death from electrical pollution, the sun and underground watercourses. Sources which can be avoided.)

Lin, J. C. (ed.) (1994). *Advances in electromagnetic fields in living systems.* Plenum.

Maes, W. (1995). *Stress durch Strom und Strahlung . . . und Gifte, Gase, Luftschadstoffe, Pilze, Fasern, Staub. Schriftenreihe Gesundes Wohnen,* 2nd edn. Institut für Baubiologie und Oekologie, Neubeuern. (Stress from electric current and radiation . . . and poisons, gases, air pollutants, fungi, fibres, dust.)

Monroe, R. A. (1971). *Journeys out of the body.* Doubleday.

Monroe, R. A. (1986). *Far Journeys.* Souvenir.

Morgan, M. (1994). *Mutant Message Down Under.* Thorsons.

Neitzke, H. P. *et al.* (1992). *Risiko Elektrosmog? Auswirkungen elektromagnetischer Felder auf Gesundheit und Umwelt.* Birkhäuser Verlag, Basel. (Risk from electrical pollution? Electromagnetic fields, their effects on health and the environment.)

Nimtz, G. and Mücker, S. (1994). *Elektrosmog.* Meyers Forum.

Ostrander, S. and Schroeder, L. (1993). *Cosmic Memory*. Simon and Schuster.

Patterson, M. A. (1986). *Hooked? NET, the new approach to drug cure*. Faber.

Payne, M. (1992). *Super Health*. Thorsons.

Perez, R. (ed.) (1995). *Handbook of EMC*. Academic Press.

Puharich, A. (1974). *Uri*. W. H. Allen.

Roney Dougal, S. (1993). *Where Science and Magic Meet*. Element.

Rose, W.-D. (1987). *Elektrostress, ein Ratgeber zum Schutz vor Gesundheitsschäden durch elektrotechnische Geräte*. Käsel-Verlag, Munich. (Electrical stress: advice on how to protect your health against electrical devices.)

Rose, W.-D. (1994). *Elektrosmog, Elektrostress. Strahlung in unserem Alltag und was wir dagegen tun können. Ein Ratgeber*, 2nd edn. KiWi. (Pollution and stress from electricity. Our daily radiation and what we can do against it.)

Rose, W.-D. (1996). *Ich stehe unter Strom. Krank durch Elektrosmog, Erfahrungen, Beispiele, Ratschläge*. KiWi. (I am at high voltage. Ill from electrical pollution. Experience, examples, advice.)

Santini, R. (1995). *Notre Santé face aux Champs Électriques et Magnétiques*. Sully, France.

Shallis, M. (1988). *The Electric Shock Book*. Souvenir Press.

Smith, C. W. and Best, S. (1989). *Electromagnetic Man*. Dent.

Steinig, H. (1994). *Elektrosmog, der unsichtbare Krankmacher*. Herder, Freiburg im Breisgau. (Electrical pollution, the invisible source of illness.)

Steneck, N. (1984). *The Microwave Debate*. MIT Press.

Sugarman, E. (1992). *Warning: the electricity around you may be dangerous to your health. How to protect yourself from magnetic fields*. Simon and Schuster.

Ulmer, G. A. (1994). *Krank durch Wellen und Elektrosmog? Wie gross ist die unsichtbare Gefahr?* G. A. Ulmer Verlag, Tuningen. (Ill because of waves and electrical pollution. How big is the invisible danger?)

Watson, L. (1974). *The Romeo Error*. Hodder and Stoughton.

Wohlfeil, G. J. (1995). *Gesund wohnen, gesund schlafen. Elektrosmog und Wohngifte vermeiden*. W. Jopp Verlag, Wiesbaden. (Living and sleeping healthily. Avoidance of electrical pollution and poisons in the house.)

World Health Organization (WHO) (1981). *Environmental Health Criteria (EHC) No. 16, Radiofrequency and Microwaves*. Jointly sponsored by WHO, UNEP and IRPA.

WHO (1984). *EHC No. 35, Extremely low-frequency fields*. Jointly sponsored by WHO, UNEP and IRPA.

WHO (1987). *EHC No. 69, Magnetic Fields*. Jointly sponsored by WHO, UNEP and IRPA.

WHO (1989). *European Series No 25, Nonionizing radiation Protection*, Suess, M. J. and Benwell-Morrison, D. (eds). Copenhagen.

WHO (1993). *EHC No. 137, Electromagnetic Fields (600 Hz to 300 GHz)*. Jointly sponsored by WHO, UNEP and IRPA.

Periodicals concerned with life

1. *Electromagnetics and VDU News*, reports on nonionizing radiation. Simon Best, PO Box 25, Liphook, Hants GU30 7SSE.

2. *The Steady Signal, Electromagnetics and Anomalies*. Albert Budden, 17 Brook Road South, Brentford, Middlesex. (Largely concerned with UFOs, unidentified flying objects.)

3. *Powerwatch Network*. Quarterly from Alasdair Philips, 2 Tower Road, Sutton, Ely, Cambs.

4. *Electrical Sensitivity News*. Six times yearly from PO Box 4146, Prescott, Arizona, 86302 USA.

5. *Network News*. Quarterly from the EMR Alliance, 410 West 53rd Street, Suite 402, New York, NY 10019, USA.

6. *Elektrosmog Report*. Monthly (in German) from Thomas Dersee, Strahlentelex, Rauxeler Weg 6, D-13507, BERLIN, Germany.

7. *EMRAA NEWS* (Electromagnetic Radiation Alliance of Australia). Sutherland Shire Environment Centre, PO Box 589, Sutherland 2232, Australia. (Quarterly)

For recent information on mind machines write to: Michael Hutchison, PO Box 2659, SAUSALITO, CA 94966-9998, USA.

Support or educational groups for electrically sensitive people

Some groups sell field meters, antistatic mats and audiotapes. All provide information. Most also sell literature.

USA

Cathy Bergman

EMR ALLIANCE
410 W. 53rd Street, Suite 402, New York, NY 10019 (quarterly newsletter)
tel. (212) 554-4073, fax. (212) 977-5541

Cellular Phone Taskforce
Arthur Firstenberg, PO Box 100404, Brooklyn, NY 11210

Ergotec Association Inc.
Bert Dumpé, PO Box 9571, Arlington, Virginia 22219
tel. and fax (703) 516-4576

IBE (International Institute for Bau-Biologie and Ecology Inc.)
Helmut Ziehe, Box 387, Clearwater, Florida FL 3615 (IBE Newsletter every two months)

Lucinda Grant
Electrical Sensitivity Network, PO Box 4146, Prescott AZ, 86302 (Newsletter 6 times yearly)

S.T.A.T.E. Foundation
Sensitive To A Toxic Environment Inc. (Mainly concerned with allergies and multiple chemical sensitivities)
4 Hazel Court, West Seneca, New York 14224
tel. (716) 675-1164

Australia

The EMR Safety Network
Convenor: Betty Venables
216 President Avenue, Miranda, NSW 2228
tel. (02) 95 40 3936

Radiation Awareness Network
Sarah Benson
c/o Croydon Post Office, Croydon, VIC. 3136

EMRAA (Electromagnetic Radiation Alliance of Australia)
Sutherland Shire Environment Centre Inc. PO Box 589, Sutherland, NSW 2232

Austria

IGEF (Internationale Gesellschaft für Elektrosmog-Forschung)
(International Electrosmog Research)
Pramaweg 45, A-6353 Going am Wilden Kaiser

Belgium

FEE – EUROSANTE
30 rue J. B. Colyns, 1060 Bruxelles

TESLABEL
Jean Delcoigne (President), rue Beauregard 12, B-7910 ARC Wattripont-Anvaing
tel. 003269 455980

Denmark

EBD, the Danish Association for the Electromagnetically Hypersensitive
Mrs Aase Thomassen
Lunden 1, Alum, 8900 Randers
tel. 86 46 61 14

France

AEVICEM (Association Européenne d'Aide aux Victimes des Champs Électromagnétiques)
c/o Martine Charpenet, 21 430 Liernais, Bourgogne

Germany

W. Bettenhausen
Arbeitskreis für Elektrosensible e.V., Alleestrasse 135, D-44793 Bochum-Stahlhausen

Bundesverband gegen Elektrosmog e.V.i.G.

(National organization for all Germany)
Frau Friederike Kochem, Klosterstrasse 9, 65391 Lorch
tel. and fax. 0234 47 3585

ECOLOG – Institut für Sozialökologische Forschung und Bildung GmbH
Nieschlagstraße 26, D-30449 Hanover (Quarterly EMF MONITOR and occasional publications)
tel. 0049 511 924 5646, fax 0049 511 924 5648

Heinz Steinig
(Author of *Elektrosmog der Unsichtbare Krankmacher*, 1994)
Arbeitsgemeinschaft Leiden unter Spannung, Badener Strasse 23, D-65824 Schwalbach

Bürgerinitiative Elektrosmog
Walter Ruck Robert-Schumann-Weg 4, D-23556 Lübeck

IGEF Deutschland (head office, see Austria)
D-72813 St Johann, Germany.

Institut für Baubiologie und Ökologie
D-83115 Neubeuern
(Occasional periodical *Wohnung und Gesundheit*, also books)
tel. 0049 8035

Institut für Mensch und Natur e.V.
Dipl-Phys. Claudia Wostheinrich
Dipl-Biolog Christof Grundmann
Obere Strasse 41, D-27283 Verden/Aller
tel. 0049 4231 81928, fax. 0049 4231 82141

KATALYSE e.V. Institut für angewandte Umweltforschung
(Environmental research institute)
Marsiliusstr. 11, D-50937 KÖLN
(Publications include quarterly *Nachrichten*: also book *Elektrosmog*, 3rd edn, 1995)

Selbsthilfeverein für Elektrosensible e.V.
im Gesundheitshaus der Stadt München, Zimmer U3
Dachauerstr. 90, D-80335 München

Japan

Tetsuo Kakehi
Editor, GAUSS NETWORK (6 times yearly)
Comfort Nakahara, Higashi Ymato-Shi
Tokyo 207, Japan
tel. (01181) 425-65-7478, fax. (01181) 425-64-8664

Norway

Foreningen for el-overfolsomme
c/o Per J. Husby, Stubbanveien, 7037 Trondheim

Sweden

FEB, Association for the Electrically and VDT-injured
Box 15126, 10465 Stockholm, Internet http://www.feb.se/
(FEB international contacts through Leif Sodergren, Daltorpsgatan 7, 41273 Göteborg)
tel. 460 712 9065

Switzerland

Schweizerische Interessengemeinschaft Baubiologie/Bauökologie
Dubsstrasse 33
8003 Zürich
tel. 01 463 4946, fax. 01 463 4849

Schweizerische Arbeitsgemeinschaft biologische Elektrotechnik (SABE)
Friedackerstr. 7, Postfach 8050, Zürich

United Kingdom

Circuit
PO Box 1UZ, Newcastle-upon-Tyne NE99 1UZ

Energy
Bill and Margaret Singleton, 70 Warwick Road, Radcliffe, Manchester M26 0WL

Powerwatch network (quarterly bulletin)
Alasdair Philips, 2 Tower Road, Sutton, Ely, CAMBS, CB6 2QA

Carshalton Ltd (ELF Monitoring Service)
Incorporating: Subtle Energy Information & Technology Institute, 14 Chilton Road, Kew, Richmond, Surrey, TW9 4HB

Chapter 15
Glossary

An asterisk (*) before a word indicates its explanation in alphabetical order elsewhere. Only some selected words are explained, that could help in understanding the texts listed at the end of the book specifying details of electromagnetism, health, etc. Ordinary conversational senses are not explained, nor are electrical terms such as amplifier, computer chip, electricity, electronics, radio receiver. Abbreviations of some organizations are listed at the end of Chapter 2.

absorption of noise Elimination of *noise by its conversion to heat in a *resistance.

AC, a.c., alternating current Mains electricity supply flowing with frequent reversals of direction, 60 times a second in USA current and 50 times in Europe. Much simpler than *DC to generate, distribute and convert to higher or lower voltages, it is more complicated to calculate because of *reactances from *inductance and *capacitance, which vary with *frequency, but in opposite senses. See tuning.

active unit A component that achieves *gain or is directional. Compare passive unit.

AD Analogue-to-digital.

ADC Analogue-to-digital converter.

AFC Automatic frequency control.

AGC (ALC, AVC) Automatic *gain (or level or volume) control of an amplifier, typically in a radio or TV receiver or recorder.

ALC Automatic level control. See AGC.

alternating current See AC.

AM *Amplitude modulation.

AMN See LISN.

ampere, amp (A) Unit of flow of electricity.

amplify Enlarge, magnify.

amplitude The height of a wave from zero to its maximum. It is half the double amplitude which is the height from plus maximum to minus maximum. Usually equivalent to voltage or power.

amplitude modulation, AM Variation of the amplitude (usually voltage) of a carrier wave so as to carry the message (speech, music or code). AM is the commonest modulation method and variants include single sideband (SSB) as well as double sideband (DSB) with suppressed carrier

AMU Antenna matching unit. See ATU.

analogue Analogue implies continuous variation, unlike *digital systems in which any continuously varying value is sampled at regular intervals to convert its smooth variation to a series of numbers (digits).

anechoic Without echos. See Chapter 2, also RAM.

ANL Automatic noise limiter, circuitry on a communications receiver, which can be switched off at will, because it can degrade the receiver's sensitivity.

antenna Modern word for aerial.

antinode A crest of a wave, compare node.

arcing Current flow across an air gap. See Chapter 6, Electrostatic discharge.

ARRL American Radio Relay League, the association which links north American radio amateurs.

astable Not stable, e.g. an oscillator.

ASTU Antenna system tuning unit. See ATU.

asymmetric current See common-mode current.

ATC Air traffic control.

ATE Automatic testing equipment, e.g. for microwave ovens coming off a production line.

athermal effect Any effect which does not involve heating.

attenuate Reduce, decrease, weaken, the opposite of *amplify.

ATU Antenna tuning unit (also *AMU or *ASTU).

audio Sound (what is audible).

aural To do with hearing.

automobile interference The voluntary standard adopted by US car builders involves a resistor built into each sparking plug, sometimes also with a high-resistance cable to the plug. But the car's alternator, electric motors, windscreen wiper, turn indicator and instruments may also generate interference. See Chapter 8.

AVC Automatic volume control. See AGC.

avionics Aviation electronics.

AVL Automatic Vehicle Location, often possible with *land mobile radio.

back EMF A *voltage opposing the normal flow of current but in particular opposing any variation of a current through a coil. See inductance.

backshell A shielded connecting plug at a cable end, joining the cable to another backshell or to the shield of an appliance. It may also ground a *shield to a cable or harness.

balancing Pairing each outflowing signal wire with its associated return wire, so that any interference appears equally in both.

balun (BALanced to UNbalanced) also called neutralizing transformer, balanced converter, bazooka, coupled inductor pair, or centre-tapped transformer. A device resembling a common-mode choke. A coaxial cable is normally unbalanced. A balun connects a coax to a balanced two-wire system.

bandwidth, or frequency range Spread of *frequencies.

BARTG The British Amateur Radio Teledata (formerly Teleprinter) Group. See packet radio.

beacon A radio help to sea or air navigation. Each beacon broadcasts its own easily recognizable repetitive signal at a fixed *frequency and at fixed times. Direction-finding apparatus on the ship or aircraft helps it. Some 3000 years ago, the first beacon known to history was a 900-km-long line of hilltop fires which told Queen Clytemnestra that Troy was taken and that her husband was returning.

beat frequency oscillator, BFO Circuitry in a radio receiver for making Morse code audible.

binary system A numbering system that relies only on two digits, '0' and '1', normal for digital computers.

bistable Description of an ability to be in one of two stable states, each of them indefinitely maintainable.

bit Binary digIT, the smallest unit of computer storage, capable only of indicating a 0 or a 1. Interference may make it change state from 0 to 1 or from 1 to 0.

blanker, noise blanker or eliminator Circuitry on a communications receiver which comes into action when the *noise from ignition systems, radar pulses, etc., becomes excessive. A circuit cuts out the radio set's reception while the noise is high. Reception is possible only between noise pulses but since the pulses do not occupy more than 20% of the time, good reception is normally assured. Rotating radar search antennas may have sector blanking in which perhaps 60° of their 360° travel may be regularly cut off for some reason.

board zoning See partitioning.

bolometer, thermal detector An instrument sensitive to a wide range of frequencies which makes use of the change in resistance of a thin conductor caused by the heating effect of the EM radiation. It can be used for measuring the strength of an EM field.

bonding Forming good electrical connections of low electrical *resistance by welding, soldering, brazing, riveting, bolting, etc.

braid Copper or copper-plated steel wire knitted to form a sleeve, to *shield a cable.

braid breaker A device that reduces stray currents travelling along *braid. It may be a common-mode choke as explained in Chapter 3, Power supplies, Fig. 3.1.

breakthrough Interference caused by a legitimate transmission, occasionally by an amateur transmitter disturbing a receiver which may have imperfect immunity. Amateurs are expected to help, and do try to avoid such troubles even though they may be transmitting quite legally. The RSGB, representing such amateurs, advises them to use the term 'breakthrough' in preference to 'interference' where any possible conflict may exist.

broadband A wide spread of *frequencies.

bus A usually thick conductor with many connections.

CAA Civil Aviation Authority.

cable harness Cables previously connected, so as to ease factory assembly of a device.

CAD Computer-aided design (or draughting).

capacitance, C Ability of a *capacitor to hold electrical charge, measured normally in microfarads (μF), nanofarads (nF), or picofarads (pF). Capacitive *reactances diminish with rising *frequency and rise with falling frequency.

capacitive Description of something that has *capacitance.

capacitor A device with *capacitance. It usually has two metal plates separated by an insulator called a dielectric. A capacitor blocks *DC but allows AC to pass, at a rate depending on its dimensions and the *frequency.

car alarms As early as 1985 car alarms could be triggered by a radio transmitter in a passing car even though these alarms were mainly operated by ultrasound. The Radiocommunications Agency will not investigate disturbances from or to car alarms. Chapters 1 and 13 mention other mobile phone problems.

carrier wave (CW) A continuous radio wave without the variations imposed by *modulation on to a radio wave that carries music, speech, etc. A carrier wave is suitable for transmitting Morse code.

catalyst A substance which speeds up a chemical reaction without itself being consumed.

cathode-ray tube An evacuated glass electronic tube usually operated at several thousand volts, to form a display for a TV or computer or other device.

CB Citizens' Band radio, originally with only one *frequency (band) but popular with long-distance drivers.

cellular radio Mobile (mainly hand-held) radio telephones which were first commercially available in the UK in 1985, and spread quickly.

Cermet Conductive ceramic used in variable *resistors.

CFL See compact fluorescent lamp.

channel A band of *frequencies used for a particular transmission.

characteristic impedance The *impedance at which a transmission line transfers its power most efficiently, without reflection or *standing wave. It is about 70 ohms for a typical coaxial cable but from 300 to 700 ohms for other lines.

choke, choking coil An iron- or ferrite-cored *inductor with high *reactance but low resistance to *DC.

circuit A ring, usually of copper wire, along which electric current can flow. It is not usually circular.

clamping A clamping voltage is the largest voltage allowed by a *transient suppressor.

clip-on choke See common-mode choke.

CM See common-mode current.

CMOS Complementary MOS (Metal oxide semiconductor). Like many *ICs it can easily be destroyed by ESD (electrostatic discharge) from the fingers.

CMRR Common-mode rejection ratio. A ratio which shows how likely it is that removal of an unwanted disturbance will also remove a wanted signal.

coaxial cable, coax A cable used for electric current at the *frequencies of radio or TV. The two conductors are concentric, the outer metal tube being separated from the inner wire by insulator.

coherent frequency A *frequency which is relatively constant, such as mains frequency (50 Hz).

common-mode choke (or ferrite-cored choke) A *ferrite ring that inhibits high-frequency interference in a conductor threaded into the ring. Wrapping the conductor round a ferrite rod is another way of excluding interference although often less efficient than a ring. A split ferrite ring may be called a clip-on choke, a snap-on or split-bead ferrite. See Chapter 3, Power supplies.

common-mode current (CM), asymmetric current Unwanted current flowing in the same direction along all conductors in a cable, including the shield over it. CM always creates interference. Compare differential-mode current.

communication(s) receiver An expensive radio receiver, commonly used for satellite reception, capable of receiving a faint, distant signal without drowning it in *noise.

commutator A cylindrical array of electrical contacts, rotating on the shaft of a generator or motor. Stationary brushes touch the moving contacts, with unavoidable sparking. In a generator the brushes draw current from the contacts but in a motor they supply current to them. The sparking produces *noise which may be eliminated by a simple noise suppression circuit using capacitors. A brushless motor, without commutator, is much less noisy.

compact fluorescent lamp, CFL, also called economy lamp Fluorescent lamps in which the glass tube is either very short or 'folded' to save space. In 1997 they were very expensive (£7 to £19), many times the cost of incandescent bulbs. They use much less power per unit of light and last longer but produce considerable RF interference. The EMC committee of the *RSGB found that the worst were 40 dB noisier than the best.

Compton effect, or scattering Release of a free electron from an atom struck by a gamma ray.

computer Although, because of the requirements of electronics, all computers work on low-voltage DC (5–25 volts), the DC is switched off and on usually at megahertz *frequencies, producing interference nearby.

connector An assembly with mating contacts for pins to enter a socket and link a cable to another cable or to other equipment. To avoid damage by electrostatic discharge to pins being touched by bare hands, the pins may be partly embedded in insulator.

converter A circuit used to change one form of information or signal to another, e.g. analogue to digital converter, AC to DC converter or voltage to frequency converter.

corona See glow discharge.

coupling Pick-up of electromagnetic energy, often *noise from one *circuit to another by *inductance, *capacitance or *conduction or all three together. One typical intentional coupling is between the coils in a *transformer.

cross polarization Receiving vertically polarized radio waves on a horizontal antenna or horizontally polarized waves on a vertical antenna. If cross polarization is avoided, better reception of radio or TV is likely.

crosstalk Everyone has heard unwanted, intelligible or unintelligible crosstalk on the telephone. But it exists elsewhere as unwanted transfer of energy from one circuit to another.

CRT *Cathode-ray tube.

cutoff frequency See waveguide.

CW Continuous wave, or *carrier wave, an unmodulated radio wave, often implying Morse code.

DA Digital-to-analogue.

dB deciBel, see Chapter 1.

dBm deciBels in relation to 1 milliwatt.

dBw deciBels in relation to 1 watt.

DC, d.c., direct current Electric current that flows in one direction only. A DC circuit is defined completely by its current, *voltage and *resistance. But *AC running through circuit suffers further obstruction from *reactances because of *inductance or *capacitance in the circuit. Together with the resistances the reactances obstruct the flow of AC.

decade About ten *frequencies (or other units).

decibel, dB The usual ratio for comparing powers or *noise or signals. See Table 1.3 at the end of Chapter 1 for dB values for powers, voltages and currents.

decoupling Elimination of a link between a source of EM interference and a victim, sometimes achieved by a *shield between them, or by short-circuiting the high-frequency energy with a *capacitor to ground.

DEF STAN Prefix to the title of a UK military standard.

demodulation In a radio receiver, the electronics that removes the *modulation from the radio wave to convert it to intelligible sound.

detector, envelope detector, decoder, demodulator The stage in a radio receiver that recovers the music, speech, etc., from the *modulations of the radio wave.

DF Direction-finding.

diathermy Heating within an animal body by electromagnetic waves for healing purposes. *Frequencies vary from 1 Hz to 2.45 GHz (microwave). Cells die at or above 43°C so overheating can be dangerous.

dielectric See insulator.

differential-mode current, symmetric current Current flowing along two conductors together, one being the 'forward' and the other the 'return' conductor. Because these two wires are close, they cancel each other's electromagnetic fields and broadcast little if any *noise. Compare common-mode current.

diffraction Bending of radio waves round obstacles. Compare refraction.

digipeater A radio *repeater station which recognizes its call sign in the address of a *packet message and automatically retransmits the message for recovery by distant stations.

digital systems Electronics in computers, microprocessors and calculators, usually based on two switch states, on and off. The switching produces interference.

DIL, DIP Dual-in-line package, the commonest type of chip package, often ceramic, having two lines of as many as 20 connecting pins.

diode Any electronic device with only two electrodes, nowadays usually a semiconductor diode. It allows current through in only one direction. It often forms part of a computer logic circuit and is used in *rectifiers. But a glass vacuum tube (thermionic valve) can be a valve diode though it might also be a triode, tetrode, pentode, etc.

dipole Any system with two distinct poles, such as a magnet, but usually meaning an antenna with a straight wire half a wavelength long, split at the middle for connection to a coaxial cable (centre-fed antenna).

direct current See DC.

directional antenna An antenna which aims in a particular direction. See beacon.

discrete Separate, the opposite of 'integrated' as in 'integrated circuit'.

discrimination See selectivity.

DM See differential mode.

Doppler effect The variation in frequency that results from a source of sound that is moving in relation to the hearer. The effect is heard when a train whistles as it passes a listener. The approaching sound has a higher frequency than the receding sound, indicating two opposite Doppler shifts. It is used in radar for finding the speed of a target, and in astronomy (red shift).

dosimetry Accurate measurement of an exposure to electromagnetic energy.

double superhet A *superhet radio receiver with two stages using different *intermediate frequencies.

downlink The radio link between the *transponder on an artificial satellite and the earth station it communicates with. The downlink frequency is always different from and often lower than the *uplink frequency.

dry joint A badly soldered joint of high electrical resistance, which may need re-soldering.

DSB Double sideband, compare SSB.

ducting Radio waves sometimes travel along ducts created by differences of temperature or humidity in the air. See Chapter 4.

dummy load An electrical unit connected to a transmitter to replace the antenna connection. It is used when an amateur wishes to check, for example, whether it is his transmitter or his antenna which is guilty of causing *interference.

E Symbol for electric field strength as opposed to 'H' for magnetic field strength.

ECG, EKG, electrocardiograph A graphical record of the voltage variations due to the heart's cycle.

EED Electro-explosive device, known to miners and quarrymen as an electric detonator.

EEG, electroencephalograph A graphical record of brain voltages.

EKG German for ECG.

electrically small (or short) In relation to the wavelength in use this generally means much less than a quarter wavelength, usually one-twentieth. A piece of metal one quarter-wavelength long can act as an *antenna.

electromagnet One of the earliest electrical devices. Its soft iron core is magnetized only when a direct current passes through the coil wound round it, but loses its magnetism when the power is cut off. It is used as an actuator for many purposes including the *relay and the *solenoid but also to lift scrap iron when hung from a crane hook.

electromagnetic, EM EM includes, in addition to RF (radio frequencies), also heat, light and ionizing waves at the top of the frequency spectrum as well as the lowest frequencies at the bottom. See Table 1.2, Chapter 1.

electrostatic discharge, ESD See Chapter 6.

electrostatic shield A *shield against electric fields.

eliminator, noise eliminator See blanker.

EMC Electromagnetic compatibility (Chapter 1).

EME, eme (1) *Moonbounce (earth–moon–earth).
(2) Electromagnetic environment.

EMF, emf (1) Electromotive force, potential difference, the voltage with no current flowing.
(2) Electromagnetic field.

EMG Electromyography for investigating muscle function.

EMI Electromagnetic interference.

emission EMC legal parlance considers emissions either as 'radiated' or 'conducted' (brought in on a cable). There are differences in detail between Germany (VDE) and the USA (FCC) but most are based on CISPR standards.

EMP Electromagnetic pulse, see NEMP.

EN Euronorm (European standard).

EPA The US Environmental Protection Agency, more powerful than the UK Health and Safety Executive because it has much wider responsibilities.

erp, ERP Effective radiated power of a transmitter in watts. It is equal to the radiative power multiplied by the *gain of the antenna.

ESD *Electrostatic discharge.

EUT Equipment under test.

exosphere The space outside the *ionosphere, more than 800 km up.

fading Reduction in the quality of radio reception, often because of fluctuations in the paths by which radio waves reach the receiver,- whether reflected from the *ionosphere or travelling along the ground. See Chapter 13, Fig. 13.3.

farad A very large unit of *capacitance. More usual units are the microfarad (μF), nanofarad (nF), and picofarad (pF).

Faraday cage A metal or wire mesh *shield that completely encircles something to be protected from EM waves, or to prevent their exit.

far field The area farther from the source of a wave than one-sixth of the wavelength. Compare near field.

FDTD Finite-difference time domain.

feeder A down lead to or from an *antenna, such as a *coaxial cable, or a twin balanced line, or a waveguide.

feedthrough capacitor A *capacitor of which one plate is bolted or soldered to the body of a *shield forming a compartment, the other being connected to the feed wire passing through the capacitor.

ferrite bead Ferrites (ferrospinels) are sintered compressed magnetic materials of high *resistance, mainly ferric oxide. Each maker has his own secret composition. One ferrite ring (called a bead) may be held on either a straight wire or one turn of wire. At DC or low frequencies it filters out the high *frequencies of interference without loss of power. Multiple beads or multiple turns may help but either may cause troublesome *parasitic capacitance. At higher frequencies ferrites become less magnetic. See common-mode choke.

ferromagnetic materials Metals like iron, steel, nickel, cobalt and some alloys that are strongly attracted by a magnet or are easily magnetized.

FET Field-effect transistor.

filter A device connected into a *circuit to exclude a range of *frequencies, the stopband, while allowing passage to other frequencies, the passband. See Chapter 11, Filters.

flicker noise *Noise which is greatest at low *frequencies and is therefore often called 1/f noise since it increases as the frequency drops. Also called low-frequency noise, current noise, contact noise, pink noise, semiconductor noise.

flip-flop A fundamental constructional unit of computers or microprocessors, capable only of storing a '0' or a '1'. It is the basis of every register, counter or memory in a computer.

floating Description of a unit not connected to the ground (earth).

flux density Magnetic force, measured in teslas. In a magnetic *circuit it is comparable to current in an electrical circuit.

FM Frequency *modulation, now available throughout the UK. It gives clearer reception than *AM.

FOT Best working *frequency from the French, fréquence optimum de travail. It is closely related to the *MUF.

frequency, Hz Number of alternations per second. How many times per second an EM wave or anything else occurs or reverses. DC has zero frequency. See Table 1.2, Chapter 1.

frequency hopping Repeated changes of *frequency, used on military radio receivers to guarantee secrecy.

frequency modulation, FM Variation of the *frequency of the carrier wave by an amount proportional to the instantaneous value of the modulating wave.

fullerenes (buckminsterfullerenes) Many ceramic compounds of a 60-atom carbon chain (C_{60}) with a metal, such as K_3C_{60}, sometimes also doped. Some are superconducting at a relatively high temperature, e.g. $-176°C$. The molecule is believed to be shaped like the architect Buckminster Fuller's football-shaped geodetic dome.

functional interference Disturbance of a device by its own circuits.

fundamental frequency A centre *frequency.

fundamental overload Disturbance of a TV or radio receiver caused by reception of the *fundamental frequency of a nearby transmitter, possibly amateur, a common type of *breakthrough.

GaAs Gallium arsenide, a semiconductor that works many times as fast as silicon but is correspondingly expensive.

gain Magnification. For an amplifier, the power output divided by the input. For a directional *antenna the gain shows how well the antenna beam is concentrated in a desired direction.

ganged capacitors Several *capacitors mounted on a common spindle.

geomagnetic field, GMF The earth's magnetic field.

GHz Gigahertz, thousand million hertz.

giga- Prefix for thousand million.

glow discharge, corona or Townsend discharge Current flow between two electrical contacts separated by a gap filled by ionized gas. See Chapter 6, Electrostatic discharge.

GMF *Geomagnetic field, the earth's magnetic field.

GPS Global positioning system using satellites.

ground loop See loop.

ground plane A continuous metal sheet, usually copper, connected to the ground, sometimes forming one layer of a multilayer or two-layer printed circuit board.

H Magnetic field strength, measured in teslas.

hardening EMI hardening ensures that a device can resist electromagnetic interference.

hard error Permanent failure.

harmonic, or harmonic component Any *frequency which is a whole-number multiple of some *fundamental frequency. For example the third harmonic has exactly three times the frequency of the fundamental.

hazard monitor A badge worn on the clothes to show, for ionizing radiation, whether the wearer has suffered too much exposure. For nonionizing radiation such as RF or microwave, these monitors are in a much more primitive state of development, although the maker Narda has some units.

heat-sink plane A continuous metal sheet in a *PCB, sometimes multi-layer, that leads heat away from a vulnerable unit. It acts as a heat sink.

HEMT High electron mobility transistor (for microwaves).

henry, L The unit of *inductance. It is the inductance of a device in which a change of current of 1 amp per second alters the electrical pressure by 1 volt.

hertz, Hz One cycle per second, the unit of *frequency.

heterodyne principle (1) The principle that any wave can be suppressed by adding to it an otherwise similar wave that is 180° out of step (out of phase) with it. From the Greek for 'external force'. See superhet.

(2) In a radio receiver, the mixing of two different *frequencies, of which one is received from the *antenna, to produce two new frequencies, respectively the sum of and the difference between the two original frequencies. One of the new frequencies, the wanted frequency, is processed further, the other frequency (the unwanted 'image frequency') is suppressed.

HF, hf High frequency. See Table 1.1.

hot spot (1) An electromagnetic hot spot is a place where several EM effects unite to create an area of high and very variable EM field.

(2) An area of the human body where the *SAR is likely to

be 5–10 times as much as elsewhere, especially in the range 30–400 MHz for the knee, ankle, elbow or wrist.

HTS High-temperature superconducting.

hyperthermia Heating of flesh or other tissue above the level at which the heat can be removed naturally. As a medical treatment it involves heating often by microwave for 30 minutes or more to perhaps 41°C. The word may also mean 'exhaustion from excess heat'. Compare below.

hypothermia Physical suffering caused by cold. Compare above.

IC *Integrated circuit.

IF *Intermediate frequency in a *superhet receiver.

image frequency The unwanted *frequency in a *superhet. It is the local oscillator frequency plus or minus the intermediate frequency.

immunity threshold The tolerable lower limit of something above which people or equipment may suffer from it. The IEC and others have defined immunity thresholds of electric fields for:
Class 1 units in a field at 1 V/m (volt per metre),
Class 2 at 3 V/m,
Class 3 at 10 V/m.

impedance The complex ratio of *AC voltage to AC current in a circuit. Like *resistance it is measured in ohms but unlike resistance it varies with *frequency. It depends on the *reactances of *inductance and *capacitance which also vary with frequency, although in opposite senses depending on whether the current leads or lags the voltage. Mathematically impedance (in ohms) $=\sqrt{(\text{resistance}^2+\text{reactance}^2)}$.

impedance matching Ensuring that the *impedances of neighbouring *AC circuits or appliances are the same. See Chapter 8.

impulse noise *Noise caused by a single momentary disturbance, unlike *white noise.

inductance (L, henrys) The typical inductance comes from a wire coil and the magnetic field of a current flowing through it. The coil provides little or no *resistance to *DC but an appreciable obstacle (*impedance) to *AC. This simple statement is all one needs to remember but the subject is often complicated by unnecessary discussions of mutual induction and self induction, explained below:

If a varying or alternating current flows through a coil that is near another circuit, the current variations in it, by mutual induction, will induce currents and voltages in the second circuit. Within the first coil, the variations in current through self induction create *back EMFs and currents smaller than, but opposed to, the variations. See henry, tuning.

induction See inductance.

induction field See near field.

inductive Concerned with *inductance.

inductor A coil or other *circuit with *inductance.

insertion loss Reduction in the voltage available to a *circuit, caused by the connection to it of an amplifier, filter, *transformer, etc. It depends on the *frequency and on neighbouring *impedances.

instability Unwanted *oscillation.

insulator A solid material that does not conduct electricity. A dielectric also is a nonconductor but may be a liquid or a gas.

integrated circuit, IC, microcircuit A semiconductor device or *circuit, usually of silicon and smaller than 1 mm.

interference, noise Disturbance in reception of radio or TV, usually from outside the receiver. Where any possibility of conflict exists, the Radio Society of Great Britain prefers the term 'breakthrough'. Except for motor ignition noise, which is loudest at high *frequencies, most noise decreases as frequency rises.

intermediate frequency, IF In a *superhet, the *frequency obtained by combining the input from the *antenna with that from the *local oscillator, usually at a fixed lower frequency than the input, originally because amplification is easier at IF.

intermodulation noise *Noise emitted not only from the *harmonics but also from the sums and differences of the frequencies of an input signal. The number of such intermodulation products can be very large. See heterodyne principle, rusty-bolt effect.

intersystem Between different systems or appliances.

intrasystem Within one system or appliance.

intrinsic noise *Noise generated within a *circuit, unlike external noise entering from outside.

inverter (1) A device that converts *DC to *AC, the opposite of a *rectifier.

(2) A NOT gate in a computer. It reverses an input by converting a 1 input to a 0 or a 0 input to a 1.

ion An atom or molecule which by the loss or gain of one or more electrons acquires a positive or negative charge. This enables it to carry electric current in an electrolyte or elsewhere.

ionizing radiation Radiations like those from a nuclear explosion. The boundary between ionizing and nonionizing radiation (NIR) is in the ultraviolet range at a *frequency above 3000 terahertz.

ionosphere A spherical volume round the Earth above its atmosphere, from about 60–500 km up, in which there are high concentrations of ions and other electrically-charged particles. In daytime it has three layers, the D-layer from 60–90 km up, the E-layer from 100–150 km and the F-layer from 160–500 km. The E- and D-layers disappear at night. All layers are ionized and thus created by the sun's ultraviolet radiation and by particles sent from the sun so the strongest ionization is at noon or soon after. The ionosphere is always fluctuating, in continuous turbulent motion.

ISM Industrial, scientific and medical equipment, restricted to special *frequencies, e.g. 13.56 MHz or multiples of it such as 27.1 or 54.2 MHz, in part to restrict the heavy interference it can cause.

isotropic Having the same properties in every direction.

IT Information technology.

jitter Erratic motion of the picture on a *CRT screen, sometimes caused by an outside disturbing magnetic field, possibly a desk lamp or a microfilm reader or a nearby transformer station.

Johnson noise See thermal noise.

kilo- Prefix for 1000 times.

laminate A product for a printed-circuit board made by bonding together two or more layers of *base material.

LAN Local area network.

land, track A conductor on a *printed-circuit board (PCB).

land mobile radio (LMR or PMR) Radio telephones run by organizations at *frequencies allocated to them by the UK Department of Trade & Industry's Radiocommunication Agency. The organization can thus keep in touch with outlying vehicles and staff, sometimes using *automatic vehicle location. LMR is used by hospitals, police, other emergency services, local authorities, business and industry. It may include paging systems, and often allows speech in only one direction at a time, peppered with 'over to you'. In thinly populated areas like Alaska, more economic methods are probably satellite radio or *meteor-burst communication. But in Europe in 1994 there were some 4.4 million users of LMR.

laser Light amplification by stimulated emission of radiation.

latch A circuit that 'freezes' a display while a counter continues to count.

LC Inductance-capacitance.

LCD Liquid-crystal diode (display).

LCR Inductance-capacitance-resistance.

LED Light-emitting diode.

LIDAR Light detection and ranging (*radar at the frequencies of visible light).

light-emitting diode (LED) The commonest *opto-electronic device, known to most people as a small red indicator light, sometimes intermittent. It is durable, cheap, and uses very little current, often only 10 mA. Compare liquid-crystal display, photodiode.

linear electronics The electronics of *analogue systems, including audio, hi-fi and TV as opposed to *digital systems.

linearity Uniform, straight-line variation, with output directly proportional to input. Compare nonlinear.

liquid-crystal display (LCD) An illuminated display that uses even less current than the *LED, about 5 μA from a 3 V supply. It is visible in bright sunlight unlike the LED, but is more fragile, having glass plates in its construction.

LISN (or AMN, artificial mains network) Line impedance stabilization network. An inexpensive 'black box' circuit that provides the stable voltage and impedance needed in testing any appliance because every mains *impedance changes as the *frequency changes. The mains supply is also affected by what is connected to it both near and far. So before any testing begins, a suitable LISN is connected to each phase between the equipment under test (EUT) and the mains. Apart from providing a known impedance it should exclude from the EUT any *noise brought by the mains. It can also inject an RF signal on to a power line that supplies the EUT.

LMR *Land mobile radio.

LNB, low-noise block A down converter on a dish *antenna that receives faint signals at (say) 12 GHz from a satellite and reduces their *frequency to a more manageable level around 1 GHz.

local oscillator In a *superhet radio receiver a normally tunable *oscillator which, combined with the input from the *antenna in the *mixer, helps to create a fixed *intermediate frequency.

loop, ground loop A circuit with connections to ground (earth) forming areas large enough to pick up electromagnetic interference. Loops even in printed-circuit boards are undesirable and in design are eliminated or made smaller. *Stray *capacitances resulting from high *frequencies can make any metal loops in a building undesirable, including water or gas pipes or the steel reinforcement of the building.

LOS Line of sight.

lossless An ideal telephone or other signalling line with no series resistance or leakage has no loss.

lossy Description of a telephone or other signalling line with high *impedance, especially *resistance.

LSB Lower sideband. See SSB.

LSI Large-scale integration, with 1000–5000 components per chip. Compare VLSI.

lumping Adding together the *resistances of the *capacitors, *inductors and *resistors in a *circuit and treating this sum as its only resistance, to simplify calculations.

magnetism Magnetism derives from the spin of the electron of a magnetic substance round its atom. Atoms with two electrons spinning round them in opposite directions are not magnetic, the two electron spins cancel each other. Atoms with a single electron, however, have a 'directional moment' because the spin of the unpaired electron creates the magnetism.

magnetophosphenes Flickering lights induced in the consciousness by magnetic fields stronger than 2 mT and applied at *frequencies above 10 Hz. The magnetic field generates electric currents in the eye (R. Santini, 1995). Electrophosphenes also exist.

mains-signalling Using the power-supply wires to carry messages, for example for the reading of meters, usually at *frequencies between 3 kHz and 148.5 kHz (BS EN 50065-1:1992), a possible source of *interference for any device connected to the mains.

man-made noise *Noise of human origin such as car ignitions, electric motors, neon signs, fluorescent lights, etc.

matching See impedance matching, Chapter 8.

maximum hold A feature of a meter which measures, e.g., magnetic fields, and records the highest value traversed. The maximum so recorded is not a peak but the highest mean value.

MBC *Meteor-burst communication.

mega- Prefix for million.

meteor-burst communication, MBC Automatic, computer-operated communications, based on reflections from the hundreds of tiny meteor trails available. It is unsuitable for speech but appropriate for data transmission. See Chapter 4.

MHz Megahertz, million hertz.

micro- Prefix for millionth.

micron (µm, 0.001 mm) Convenient word, used by engineers, for a millionth of a metre. The 'correct' word for 0.001 mm is 'micrometre', disliked by engineers because it also means an engineering gauge.

microphonics *Noise in an audio system, often caused by flexing of a cable and consequent rubbing of its insulation against its conductors or shield.

microprocessor A complete CPU (central processing unit) of a computer on one chip.

microwave Frequencies between about 1 GHz and 300 GHz but other definitions exist. See Table 1.2, Chapter 1.

milli- Prefix for thousandth.

millimetre waves A wavelength of 1 mm is at a frequency of 300 GHz.

mixer, frequency changer In a transmitter or receiver, the stage at which a signal *frequency combines with the output of a *local oscillator to form sum or difference frequencies. See also intermediate frequency.

MMIC Monolithic microwave *integrated circuit.

MMW Millimetre waves.

MMWIC Microwave and millimetre wave integrated circuit.

mobile phone Term used in Europe for *cellular radio. In the USA it implies *transceivers installed in cars.

modulation For an electromagnetic wave to carry speech or music or any other intelligible signal, the wave must suffer changes (modulation) to its power, frequency or phase in proportion to the magnitude or frequency of the signal.

moonbounce (eme, earth–moon–earth) Communication between distant amateur radio stations by radio-wave echo from the moon using sophisticated, expensive transmitters and receivers and enormous patience.

MOS Metal oxide semiconductor.

motor-generator set An AC mains-driven motor on the same shaft as an electric generator for DC or high-frequency AC.

moxing Producing an *intermediate frequency by 'beating' the incoming signal with a locally produced oscillation in a *mixer.

MSI Medium-scale integration, with 10 to 100 components per chip.

MUF Maximum usable *frequency. At any particular place and time, the highest frequency practicable for sending RF waves to the *ionosphere for reflection to Earth. In practice the frequency used for transmission is about 80% of the MUF.

multipath reception (See Chapter 4) Reception of TV or radio can be distorted by *interference between two or more phases of the radio wave. The two phases are not in step because one phase comes direct to the receiver, other(s) being reflected off buildings, etc. Being delayed they are out of phase with the direct signal and distort it. At 180° out of phase they would cancel each other.

MW Microwave.

nano- Prefix for thousand-millionth.

nanon Thousand-millionth of 1 m, 0.001 μm.

narrow band Concerned with few *frequencies.

near field, induction field The area nearer the source of a broadcast than one-sixth of the wavelength. Compare far field.

negative feedback A way of improving the stability of a *circuit. An electrical connection from the output back to the input is so arranged that any change in output automatically reduces whatever causes the change, unlike positive feedback.

NEMP Nuclear electromagnetic pulse, usually from an atom bomb or related device. A NEMP at 100 km above ground could smash any microelectronic device that receives it, ruining communications.

NIR Nonionizing radiation.

node A point of zero movement in a wave. Compare antinode.

noise Disturbance at a lower level (less loud) than *interference. Unwanted electric currents, usually *broadband, that spoil radio sound or TV pictures and other signals. Noise has to be reduced to achieve good reception, and in *digital systems like computers to prevent triggering at the wrong level. Two main types of noise are:

(1) internal noise generated in the receiver's semiconductors and resistances,

(2) external noise received from the antenna or the mains power supply, either man-made or natural. Noise may often be in part excluded or reduced by *filters or other electronic devices that improve *selectivity. At low or medium *frequencies, noise received on the antenna is much larger than internal noise. But at VHF (30–300 MHz) and higher frequencies most of the noise originates within the receiver.

noise blanker See blanker.

noise factor Like the SNR, signal-to-noise ratio, the noise factor is also a ratio, thus:

$$\text{input SNR} \div \text{output SNR}$$

It shows how much the input signal has been degraded by the internal *noise in the receiver. At *frequencies above 30 MHz the noise from the antenna is much smaller than internal noise, so a low noise factor is needed at such frequencies.

noise suppression Preventing the transfer of electronic *noise by a relatively simple adjustment such as a *filter to a noisy circuit or device.

nonionizing radiation (NIR) EM waves with a *frequency below ultraviolet, sometimes including acoustic or other non-EM waves.

nonlinear Description of a variation which is disproportionate to input, with a great increase or decrease of output for a small change in input. Many amplifiers are intentionally nonlinear. Compare linearity.

OATS Open area test site, sometimes called an open-field test site. The latter could exclude a flat roof.

ohm (Ω) The unit of electrical *resistance, also used for measuring *impedance.

op-amp, operational amplifier A versatile general-purpose amplifier circuit, popular since the 1930s since it is good at maths. It can be used to add, subtract, divide, multiply, differentiate or integrate and can be bought cheaply as an *integrated circuit.

open circuit A *circuit in which no current flows.

optical fibre Extremely fine glass or plastic fibre which transmits a ray of light over a long distance with little or no reduction in strength. It cannot suffer *EMI nor does it emit any.

optocoupler An *optoisolator.

optoelectronics All semiconductors are sensitive to light, some more than others. Optoelectronics is the use of *light-emitting diodes (LED), *liquid-crystal displays, and specially photosensitive (light-sensitive) semiconductors in devices like the *photoconductive cell (photoresistor), *photodiode, *phototransistor, or *photosensitive integrated circuit. Some of them combine with each other to form an *optoisolator to send information in the form of light along an *optical fibre.

optoisolator A fast *optoelectronic unit consisting of a light emitter and a photosensitive device. In the presence of a *circuit with many transients and high voltages which could destroy sensitive ICs, it eliminates this danger by substituting optical transfer of information in place of the electrical transfer. It is more efficient than an isolating transformer but probably cannot transmit high currents.

oscillation Rapid reversals of current in a *circuit usually at radio *frequency, often accentuated by a *resonant circuit.

oscillator An amplifier providing its own input by positive feedback from output to input circuits at some definite frequency. It should have good stability in both frequency and amplitude.

OTHR Over the horizon *radar.

package An encapsulation in ceramic or plastic that protects an integrated circuit (IC). When inserted into a socket, its electrical pins unite the IC with the outside world.

packaging The details of a *package.

packet radio Radio messages sent out in short predetermined lengths under agreed protocols (conventions) to suit the various possibilities for start, stop, repeat, error correction, etc. Helped by *repeaters and the *transponders of amateur (or other) satellites, such messages travel long distances. The Internet, though in a different (wired) medium, also uses packet switching.

paging Using an extremely small, inexpensive, radio receiver (pager) providing messages to ask someone to telephone or appear in person.

PAMR, PMR Public access mobile radio. See land mobile radio.

Pandora Code name for a research project in the USA from 1965–71 to find whether the health of monkeys could be affected by microwaves similar to those suffered by the US embassy employees in Moscow. Pandora was renamed Bizarre at one time.

parasitic (stray) Every *AC *circuit as its *frequency rises acquires stray *reactances of two types, capacitive and inductive. Stray *capacitance may come merely from two parallel wires and can allow AC to pass. Stray *inductance may come from a mere bend in a wire. See self resonance. DC circuits being without frequency have no reactance, only *resistance.

partitioning (zoning) On a printed-circuit board (PCB) the placing of fast-switching circuits together and away from slower and sensitive circuits. Fast logic, clock oscillators and bus drivers should be together and close to the edge connectors of the PCB. Slower logic and memory can be furthest from the edge connectors because they can tolerate longer wiring without resulting *noise or *crosstalk.

passive unit A component such as a length of wire, having *resistance, *capacitance or *inductance or any mixture of them. Compare active unit.

PCB *Printed-circuit board.

PD Potential difference, see EMF.

permeability A magnetic property. *Ferromagnetic materials are highly permeable.

phase Different parts of an electromagnetic wave are its phases. 'Out of phase' means they are out of step, interfering with each other. The amount out of step can be measured in degrees or radians, etc. If two parts are 180° out of phase they cancel each other completely.

photoconductive cell or photoresistor A light-detecting *optoelectronic device, usually based on cadmium sulphide (CdS). Because it is sensitive to visible light it is used in camera exposure meters. In bright sunlight the electrical resistance of a CdS photoresistor drops to 150 Ω or less from its 'dark' resistance of 10 MΩ. Photoresistors do not respond quickly so are unsuitable for rapid response.

photo-Darlington Two *phototransistors combined to achieve very high amplification. It is used with a microprocessor to count objects, etc.

photodiode A semiconductor *diode that responds to light or infrared. Its typical current in the dark is only 1.5 nA. See phototransistor, light-emitting diode.

photoresistor A *photoconductive cell.

photosensitive IC An integrated circuit which is an advance on the *photo-Darlington, both in cheapness and convenience. It can switch very quickly from on to off and back and is often used in a light-actuated switch.

photosynthesis Synthesis (creation) by plants of chemical compounds in sunlight, especially of carbohydrates from carbon dioxide and water, simultaneously releasing oxygen.

phototherapy Healing by light.

phototransistor A light-sensitive *transistor amplifier about 200 times more sensitive than a *photodiode, so it is more popular.

photovoltaic cell A solar cell. In a space vehicle or calculator it converts sunlight into electrical energy for sending messages to earth or doing calculations.

pico- Prefix for one-million-millionth.

piezoelectricity Either generation of electricity by certain crystals (quartz, etc.) when they are stretched or compressed, or their stressing by an electrical signal. This reversible effect is used by many sensors such as microphones, gramophone cartridges, sonar pingers, seismic detectors, crystal-controlled oscillators and others that interchange electrical with mechanical energy.

pigtail A connection by wire only, from a cable to a ground (earth) or to an appliance shield at a break in the cable shield. A much better although expensive solution is to use a *backshell to connect the cable shield to the appliance shield for 360° all round.

pink noise *Noise like *flicker noise varying inversely with *frequency.

PIP Passive intermodulation products, the *rusty-bolt effect.

plane wave An electromagnetic wave in the *far field.

PLL Phase-locked loop.

PM Phase modulation.

PMR Private mobile radio. See land mobile radio.

polarity Of the many senses of polarity, that which most concerns *EMC is the polarity of an electromagnetic wave. A vertical *antenna sends out a vertically polarized wave, with its electric field in a vertical plane. A horizontal antenna broadcasts in a horizontal plane with horizontal polarity. By convention it is the electric field which decides the polarity. Waves with vertical polarity are best received with a vertical antenna to avoid *cross-polarization. *Mobile phones, usually with vertical 'whip' antennas, have vertical polarity. See Chapter 5 (end).

polarization See polarity.

port An electrical connection to a computer. From the viewpoint of the *EMC law there are five classes of port which may connect a unit with its environment:

1. enclosure (shield, screen),
2. AC power supply,
3. DC power supply,
4. ground (earth) connection,
5. signal, control or other connections.

positive feedback A *circuit layout in which the output reinforces the input, often resulting in *oscillation. Compare negative feedback.

potential difference, PD See EMF.

potentiometer (or loosely) variable resistor A *resistor which can be tapped at intermediate values.

power density A dosage to the animal body, measured in watts per square metre (or in the USA mW/cm^2, milliwatts per square centimetre) obtained by multiplying together the magnetic and electric field strengths. Thus

$$W/m \times A/m = watts/m^2$$
$$\text{American unit, } 1\ mW/cm^2 = 0.1\ W/m^2.$$

Compare SAR.

power factor improvement The effect of *inductive *reactance in an *AC electrical *circuit is that the current lags behind the voltage. If, however, the reactance is *capacitive, the current leads the voltage. Either of these in excess brings in the need for power factor improvement because when the current is thus not in phase with the voltage, the *DC formula:

$$\text{volts} \times \text{amps} = \text{power, watts}$$

no longer applies. With excess reactance of either type the real power output is:

$$\text{volts} \times \text{amps} \times \cos í = \text{power, watts}$$

where í is the angle of lag or lead and cos í is the power factor. Ideally cos í is equal to 1. A lagging power factor can be improved by adding *capacitance or a leading one by adding *inductance. If *capacitors are switched in or out, however, the action of switching can cause the voltage to flicker, a source of *noise. Switching in any case causes *interference. If too many capacitors are used, the voltage may rise excessively.

power flux density *Power density.

power plane or voltage plane A conducting layer in a *PCB which is held at a voltage other than ground.

preamplifier An amplifying circuit which precedes a power amplifier.

prepreg or bonding sheet or B-stage A sheet material for *PCBs impregnated with resin at an intermediate state of manufacture.

PRF Pulse repetition frequency.

printed-circuit board, PCB A *circuit (usually circuits) photographically reproduced on an insulating plastic board by the chemical removal of the thin copper foil (0.004 inch or thinner) over it. There is often also copper foil on the underside. Sometimes on a multilayer board there are additional foil layers or PCBs below or above. See ground plane.

probe A measuring or sensing device usually containing a *sensor, sometimes also an *antenna and often shaped like a pencil.

propagate Radio experts' word for how RF waves travel.

protocol Any convention agreed for telecommunication.

PWB Printed-wiring board. See printed-circuit board.

Q factor, quality factor The magnification of the output of a radio or audio appliance at *resonance.

qp, quasipeak A maximum level that corresponds to the human feeling for the maximum.

radar, radiolocation RAdio Detection And Ranging, developed in the UK between 1939 and 1945 for detecting and finding the range of enemy aircraft. Powerful pulses at *microwave *frequencies are sent out at intervals. The time taken for the echo to return is measured electronically to give an automatic indication of the range (distance).

radiation Emission of electromagnetic waves.

Radio Society of Great Britain (RSGB) The British association of 30 000 radio amateurs which represents them at governmental level and helps them to become licensed, to learn Morse code and much else.

RAM Radio- (or Radar-)absorbent material, used in a *screened or anechoic room to absorb radio echos.

RC Resistance-capacitance.

reactance Obstruction to the flow of *AC caused by *capacitance and *inductance rather than *resistance, but also measured in *ohms. It may be capacitive or inductive or both.

receptor, susceptor A victim device that suffers interference.

rectenna A combined *rectifier and *antenna.

rectifier A device which converts *AC to *DC.

refraction Bending of EM waves away from the straight, usually towards the ground, helping to increase the distance that radio waves can be sent along the ground surface. The radio horizon is then beyond the seen horizon.

regulation Voltage regulation is: maintaining a constant voltage.

relay A device in which voltage or current in one circuit switches another circuit on or off. An electromechanical relay is usually an *electromagnet operating a mechanical switch.

repeater, digipeater (1) (relay) A radio station permanently on the air, that extends the range of fixed (amateur) stations by receiving and automatically re-transmitting at a different frequency the messages that reach it. Often in the UK it acts also as a *beacon. Repeater stations are built and used both by commercial operators such as BT and by amateurs. They have to be licensed by the Radiocommunications Agency but are often operated as well as financed by radio amateurs under the *RSGB. See transponder.

(2) (repeating coil) A transformer that inductively unites the communications of two sections of a telephone line that must not be directly connected.

resistance (R, ohms) Obstruction to the flow of electric current, especially *DC. A circuit with resistance heats up in proportion to the number of ohms times the current (I) squared, so the heating effect is often called the I^2R loss, measured in watts.

resistance noise See thermal noise.

resistor A component with *resistance.

resonance A state of an *AC circuit in which the inductive and capacitive *reactances are equal, minimizing the total *reactance.

resonant circuit An *AC circuit with *inductance and *capacitance. At the resonant *frequency the two *reactances cancel each other out. *Resonance at the desired frequency is essential to optimize reception.

RF, rf Radio frequency, usually taken as any frequency from 10 kilohertz to a few gigahertz. Most interference in radio and TV receivers is at RF.

ringing A device 'rings' if it goes on producing a signal or moving after its input stops. Caused by too little damping, it can be troublesome in loudspeakers. In a low-frequency loudspeaker it is aptly called 'hangover'.

RIS Radio Investigation Service of the UK Radiocommunications Agency.

RMS, rms Root mean square, the value of an AC voltage or current, adopted to avoid the confusion created by AC variations from one maximum through zero to another maximum 100 times a second.

RSGB See Radio Society of Great Britain.

rusty-bolt effect Intermittent contact between metal units, such as fence wire with a fence post, or rainwater guttering at loose joints, producing electronic interference, often with electrolysis between different metals or metals containing impurities.

S See S-meter.

SAR Specific absorption rate, measured in watts per kg of body weight, to provide a simple way of comparing a dose hazard to an animal with one to a human. Compare power density.

saturation The greatest possible level of magnetization, after which further magnetizing current gives no further magnetism. High currents in an *inductor can saturate its iron core. Usually saturation must be avoided because it dramatically lowers the *inductance, and often simultaneously emits interfering *harmonics of the inductor current.

SAW filter Surface Acoustic Wave filter. A small, reliable, lightweight device based on *piezoelectricity that is designed for specific *frequencies in TV and radar.

screen, shield See Chapter 8.

screened room A laboratory test room shielded from erratic radio waves, in which electronic equipment can be tested for its EMC and other qualities.

sector blanking See blanker.

selectivity, discrimination For a radio receiver, its ability to choose a wanted signal and to suppress all others.

self resonance Every *capacitor has *inductance and every *inductor has *capacitance. The self-resonant frequency for either of them is the frequency at which the stray values of capacitance or inductance equal the design values of the inductor or capacitor. At or above its self-resonant frequency no capacitor or inductor can be used as designed. See parasitic.

sensitivity Of a radio receiver is measured by the smallest input voltage needed to give a stated output power at a stated signal-to-noise ratio (SNR). For example a communication receiver using amplitude modulation might have typically 2 μV (microvolts) at 300 Hz for a SNR of 20.

sensor, detector A device for measuring or merely detecting a physical quantity such as light, temperature, electric current or acidity. It may send its findings electrically to a computer. In an instrument for measuring EM field strengths the sensor must contain both an *antenna and a detector.

SGEMP System-generated electromagnetic pulse, usually in interstellar space to destroy a rocket, satellite or other target.

shield, screen See Chapter 8.

shot noise *Thermal noise.

SID (1) Sudden Ionospheric Disturbance. An abrupt increase in the ultraviolet and X-rays resulting from a solar flare suddenly increases the ionization of the D-layer over the whole sunlit side of the earth. It returns to normal only after a period varying from a few minutes to an hour. Long waves (500 kHz or higher) can propagate more easily with better reflection from the D-layer but for the higher frequencies the D-layer absorbs so much that communication may be impossible.

(2) Sudden Infant Death (cot death).

sideband A band of *frequencies above or below the nominal *carrier frequency, produced by *modulation.

signal A wanted message, not unwanted noise.

signal recovery Obtaining information from a signal partly or wholly submerged in noise. It always involves raising the *signal-to-noise ratio.

signal-to-noise ratio, SNR A number, the level of the signal plus noise divided by the level of the noise. The larger the number, the higher is the quality of the transmission. Thus for music the minimum SNR should be 60 but for a telephone conversation 30 is enough.

skin effect When *AC flows through a wire, most of the flow takes place along the outer surface, the skin, a trend intensified with rising *frequency. Thicker wire therefore does not reduce the *impedance of the wire. At 1 MHz or higher frequences, it is pointless to use wire thicker than 0.3 mm. See Chapter 9, Shields.

S-meter Signal-strength meter. It has graduations from 0 to 9 followed by '+ dB', implying the number of decibels over S9.

SNR *Signal-to-noise ratio.

sniffer An instrument for detecting field strength, mainly for finding leaks through *shields.

soft error Fault in a display or elsewhere that can be corrected by pressing the right keys.

solar wind Electrically charged particles, mainly protons and electrons with a few nuclei of heavy elements, which have been accelerated by the high temperature of the sun sufficiently to allow escape from the sun's gravitational field. The wind speed in a quiet solar period is 350–700 km/sec. When the solar wind meets the earth's magnetic field, a shock wave results which is not yet fully understood. What remains of the solar wind travels on some 3000 million miles where it cools and diffuses into galactic space.

solenoid A common actuator, based on the *electromagnet. Its soft-iron core can be pulled in or pushed out by an electric current passing through its coil.

solid-state 'Made from semiconductors' as opposed to 'vacuum-state', which implies thermionic valves and related devices.

spike A *transient with a sharp peak.

spurious Unwanted.

squelch Noise suppression on a radio set.

SSB Single Sideband, a *modulation method in which one sideband is suppressed to save spectrum space. If the carrier is also suppressed it may be re-inserted in the receiver.

standing wave A standing wave may be seen in a river below a weir where the water surface is below the general downstream level. Equivalent standing waves occur in acoustics or with electromagnetic (EM) waves in a *transmission line. An EM standing wave is normally longer than one in water, having crests at successive intervals of half a wavelength along the transmission line.

stereo, stereophonic Reproduction of sound with two loudspeakers.

stray See parasitic.

superhet, superheterodyne (originally **supersonic heterodyne**) An operating principle of almost all modern radio sets whereby the incoming RF waves are converted in a *mixer with the help of a *local oscillator to a fixed intermediate frequency that is more convenient for amplification. The mixer may be called a frequency changer. See heterodyne principle.

suppression *Noise suppression, but for transient suppression, see Chapter 10, Transients.

suppressor In cars, an electrical resistance in the cable to the spark plug silences interference from the ignition, and allows the car radio to be heard. For transient suppressors, see Chapter 10, Transients.

susceptible Prone to suffer interference.

susceptor A device that suffers interference.

SWL Short-wave (radio) listener.

SWR Standing-wave ratio.

symmetric current *Differential-mode current.

Tempest Prevention of electronic eavesdropping on US government information, largely by the use of shielded rooms, etc., officially called NACSIM 5100.

TEM wave Transverse electric and magnetic wave.

tera- Prefix for ten to the twelfth, a million million.

tesla, T The unit of magnetic flux density.

thermal detector See bolometer.

thermal noise, Johnson noise, shot noise All resistances generate this *white noise. Pure *inductances or *capacitances, if they exist, do not generate it.

thermocouple Wires of two different metals, joined at their ends, develop a voltage depending on the temperature difference between the junctions. The device can be used to measure temperature or radiant energy which it converts to electricity. It can thus be used to measure EM field strengths.

toggle To switch between two states.

top band Approximately 160 m wavelength or 2 MHz frequency.

toroid A ring or doughnut shape.

Townsend discharge See glow discharge.

trace, track, land A copper conductor on a PCB (printed-circuit board).

Trafficmaster YQ (Why queue?) An in-car radio receiver available on subscription to notify the car driver of traffic congestion, a system which started near London in 1990. See also Trafficmate.

Trafficmate plc An in-car radio receiver available on subscription to notify the car driver of traffic congestion. In 1996 such broadcasts at around 433.9 MHz (70 cm wavelength) seriously disrupted amateur TV and *repeaters, completely corrupting packet radio unless the amateur sender was quite close. The makers claimed that the transmitter output power was below 10 mW but the interference included a disruptive click every half second. See also Trafficmaster.

transceiver A radio transmitter, usually portable, which is also a receiver, used in mobile phones, *CB, *land mobile radio, etc.

transducer A *sensor which converts one sort of energy into another, e.g. electricity into sound, light or air pressure or the reverse.

transformer A relatively simple but basic *AC electrical construction consisting of two coils of wire (the primary and the secondary coils) usually wound one on top of the other on the same core of iron or other magnetic material. The number of turns in one coil divided by the number in the other is the turns ratio. The voltage put into the primary comes out of the secondary multiplied by the turns ratio. If the secondary has more turns than the primary the unit will be a step-up transformer. If there are fewer it will be a step-down transformer.

transient Isolated short-term surge or drop in voltage or current. See Chapter 10, Transients.

transistor Versatile electronic units used in radio as switches, amplifiers, etc., and in computing and telecommunications also as memory units. They have very fast switching times, often less than a nanosecond (thousand-millionth of a second). The disadvantage of fast switching is that it produces more interference than slow switching. All transistors are highly temperature-dependent and must not overheat. (Originally transfer-resistor.)

Transmatch An antenna tuning unit, similar to a *balun.

transmission line Two wires or a *coaxial cable or a *waveguide carrying radio waves, between an antenna and a receiving set or transmitter. Matching its *impedance to the impedance of the receiver or transmitter will improve the performance of the device.

transponder One type of *repeater. Its automatic electronics respond to a RF message on one *frequency by answering on a different frequency. An earth satellite usually receives the *uplink at a *frequency higher than it re-transmits for the *downlink. Transponders are used also on civil aircraft to show to fighter aircraft in pursuit that they are not hostile, also in civil life.

TRF Tuned radio frequency, used in place of *superhet operation.

triboelectric effect Transfer of electric charge between two materials which are rubbed together and then separated, leaving one positively, the other negatively charged, and creating *noise. When an insulator touches and then separates from a conductor, the separation and flow of charge leads to noise. This may happen when a coaxial cable is distorted.

tuning Usually adjusting the *capacitance or *inductance of a RF *circuit to resonate at a particular *frequency. Radio or TV receivers have to be tuned to receive a desired frequency and reduce *noise. This is possible only because luckily, capacitance and inductance behave oppositely with changing frequency. With rising frequency the inductive *reactance rises but the capacitive reactance falls. Tuning means varying one or the other until the two are equal, creating a *resonant circuit that optimizes reception or transmission.

unsquelch Removal of *squelch (switching it off).

uplink The radio link between an earth station and an artificial satellite at a frequency different from the *downlink to avoid confusion at the *transponder.

USB Upper sideband. See SSB.

vacuum-state See solid-state.

var, VAR Volt-amperes reactive.

varactor A *diode designed to act as a voltage-variable *capacitor.

varicap A variable-capacitance *diode. It is smaller and usually cheaper than a *capacitor but it must be wired with the plus electrode at the plus end.

varistor Variable-voltage *resistor, often made of silicon carbide. Its resistance drops as the applied voltage rises, sometimes providing protection against *transients.

VCO Voltage-controlled oscillator.

VDT Video display terminal.

VDU, VDT Visual display unit, monitor or screen of a computer, etc.

via A hole through a printed-circuit board that electrically connects two tracks or layers of copper foil.

volt Unit of electrical pressure.

voltage Electrical pressure in volts.

VSAT Very small aperture terminal for portable and relatively inexpensive, two-way communication by satellite.

VSWR Voltage standing-wave ratio.

VSWR meter A device to help matching an antenna to a receiver or transmitter.

waveguide A device used at *frequencies above about 1 GHz to transmit electromagnetic waves, generally through a rectangular metal pipe, instead of wires or *coaxial cable. Each waveguide can carry electromagnetic waves only above a certain frequency, called its cutoff frequency. So it may be used as a *filter to exclude lower frequencies.

wave impedance In an electromagnetic wave of a given *frequency the strength of the electric field in volts per metre divided by the strength of the magnetic field in amps per metre. Thus V/m divided by A/m equals V/A = *ohms. In the *far field it is constant at 377 ohms.

wavelength See Table 1.2, Chapter 1, also Chapter 9, Shields.

white noise Noise of uniform power over a wide range of *frequencies, unlike *impulse noise.

WLAN Wireless-based LAN (local area network).

zoning See partitioning.

Information sources

Mainly for Chapters 3 to 12. For texts concerned with health or the psyche, see Bibliography at the end of Chapter 14.

A. Technical books and periodicals
B. Other sources of information.

A1. Technical books

American Radio Relay League (1991). *Radio Frequency Interference: How to find it and fix it*. ARRL.

Barnes, J. R. (1987). *Electronic System Design & Noise Control Techniques*. Prentice-Hall.

Carr, J. J. (1983). *Practical Antenna Handbook*. TAB Books, USA.

Chatterton, P. A. and Holder, M. J. (1991). *EMC: Electromagnetic Theory to Practical Design*. John Wiley.

Davies, E. R. (1993). *Electronics, Noise and Signal Recovery*. Academic Press.

Degauque, P. and Hamelin, J. (eds) (1993). *Electromagnetic Compatibility* (translated from French). Oxford.

Encyclopedia of Earth System Science, 4 vols. (1992). Academic Press.

Fish, P. J. (1993). *Electronic Noise and Low-Noise Design*. Macmillan.

Goedbloed, J. (1992). *Electromagnetic Compatibility* (translated from the Dutch edition of 1990). Prentice-Hall.

Greason, W. D. (1992). *Electrostatic Discharge in Electronics*. Research Studies Press and John Wiley.

Hall, M. P. M. and Barclay, R. L. W. (eds) (1989). *Radiowave Propagation*. Peter Peregrinus on behalf of the IEE.

Huschka, M. (1990). *Multilayer Bonding Guide* (translated from Einführung in die Multilayer Presstechnik). Technical Reference Publications.

Keiser, B. (1987). *Principles of Electromagnetic Compatibility*, 3rd edn. Artech.

Law, Jr, P. E. (1987). *Shipboard Electromagnetics*. Artech.

Lee, J. G. (1991). *Introduction to Radio Wave Propagation*. Bernard Babani.

Lenk, J. D. (1992). *Lenk's RF Handbook, Trouble Shooting*. McGraw-Hill.

Mardiguian, M. (1992). *Controlling Radiated Emissions by Design*. van Nostrand Reinhold.

Marshman, C. (1995). *Guide to the EMC directive, 89/336/EC*, 2nd edn. EPA Press, Wendens Ambo, Saffron Walden.

Mills, J. P. (1993). *Electromagnetic Interference Reduction in Electronic Systems*. PTR Prentice-Hall Inc.

Moell, J. D. and Curlee, T. N. (1995). *Transmitter Hunting, Radio Direction-finding Simplified*. McGraw Hill.

Molyneux-Child, J. W. and Hamilton, S. (1992). *RFI/EMI Shielding Materials, a Designer's Guide*. Woodhead Publishing Ltd.

John Moore Associates Ltd. (1993). *EMC Testing in UK, a Survey of Third Party Test Houses*. John Moore.

Morgan, D. (1994). *Handbook for EMC Testing and Measurement*. Peter Peregrinus, on behalf of the Institution of Electrical Engineers.

Nelson, W. R. (1981). *Interface Handbook*. Watson-Guptill Publications.

Ott, H. W. (1988). *Noise Reduction Techniques in Electronic Systems*, 2nd edn. Wiley Interscience.

Page-Jones, R. (1992). *Radio Amateurs's Guide to EMC*. Radio Society of Great Britain.

Paul, C. R. (1992). *Introduction to Electromagnetic Compatibility*. John Wiley.

Philips, A., Mayhew, N. and Williams, T. (1994). *Living with Electricity*. Scientists for Global Responsibility.

Picquenard, A. (1974). *Radio Wave Propagation*. Macmillan (and Philips).

Scott, A. W. (1993). *Understanding Microwaves*. Wiley Interscience.

Smith, B. L. and Carpentier, M.-H. (eds) (1989). *Microwave Engineering Handbook*, 3 volumes. Chapman & Hall.

Stutzman, W. L. (1993). *Polarization in Electromagnetic Systems*. Antech.

Thornton, E. (1991). *Electrical Interference and Protection*. Ellis Horwood.

Tihanji, L. (1995). *Electromagnetic Compatibility in Power Electronics*. IEEE Press, J. K. Eckert and Butterworth-Heinemann.

Violette, J. L. N., White, D. R. J. and Violette, M. F. (1987). *Electromagnetic Compatibility Handbook*. van Nostrand Reinhold.

Walsh, R. A. (1995). *Electromechanical Design Handbook*, 2nd edn. McGraw-Hill.

Watson, J. (1991). *Mastering Electronics*, 3rd edn. Macmillan.

Williams, T. (1992). *EMC for Product Designers*. Butterworth-Heinemann.

Williams, T. (1989). *Nuclear Electromagnetic Pulse*. Electronics & Computing for Peace.

Young, E. C. (1988). *Penguin Dictionary of Electronics*, 2nd edn.

A2. Technical periodicals

The US Institute of Electrical and Electronics Engineers (IEEE) organizes an annual International Symposium on EMC and, a few months later, publishes the papers delivered there by more than a hundred authors on every conceivable EMC subject, totalling more than 500 pages every year.

The IEEE's annual International Microwave Symposium is equally bulky and informative, but about microwave.

The IEEE's Electromagnetic Compatibility Society publishes its transactions a few times a year, mainly on EMC research.

The IEE in the UK has a large library, as well as a video that teaches EMC.

Radio Communication, monthly, published by the Radio Society of Great Britain (RSGB) for its members, who are all radio amateurs. Occasional articles are 100% about EMI. But since radio amateurs are always blamed by their neighbours for interference suffered in radio or TV reception, RSGB prefers the word 'breakthrough' (by amateur broadcasters) to 'interference'.

B. Other sources of information

The Department of Trade and Industry, Waterloo Road, London SE1, has for many years provided information on EMC.

The Institution of Electrical Engineers, Savoy Place, London, has videos on EMC which are available to members, and by arrangement to others.

Many universities provide EMC studies, especially the universities of York and Hull, which cooperate to provide undergraduate and postgraduate courses in EMC and in RF communications, mostly at the Department of Electronics, Heslington, York. York University is also linked to KHBO, the Catholic University of Bruges and Oostende which specializes in teaching EMC, ESD, and shielding.

The York Electronics Centre, part of the University's Electronics Department, helps large or small firms to use electronics in their products or in manufacturing processes. Research in the department, using its screened rooms for measurement of interference, is extremely wide ranging and covers parallel processing, computer music, the singing voice, speech analysis, and transducers to help physically-disabled people.

ERA Technology Ltd, of Cleeve Road, Leatherhead, Surrey, KT22 7SA, formerly the Electrical Research Association, publishes texts on electrical matters including EMC for its members, mainly industrial organizations. Strictly speaking, its publications are not sold to non-members, but this rule may be waived for older texts.

Appendix A
Events of electromagnetic history

Year	*Discoverer/organization*	*What was achieved*
640 BC	Thales of Miletus	Static electricity by rubbing amber
Dates AD		
1100	Chinese and Arab sailors	Lodestone as compass
1746	P. van Musschenbroek	First capacitor, Leyden jar
1752	B. Franklin	Flew a kite in a thunderstorm and connected its string to a Leyden jar
1800	F. W. Herschel	Infrared radiation from the sun
1800	A. Volta	Voltaic pile, the first battery
1800	A. Volta	Electrophonic hearing
1820	H. C. Oersted	Elcctromagnetism
1820	Schweigger of Halle	Galvanometer
1823	W. Sturgeon	First electromagnet
1827	G. S. Ohm	Ohm's law
1828	J. Henry	Developed electromagnets
1831	M. Faraday	Electromagnetic induction
1831	M. Faraday	Transformer: electrolysis
1832	J. Henry	Self induction
1834	Charles Babbage	Designed a computer but did not build it
1843	S. F. B. Morse and A. N. Vail	Devised the Morse code
1856	S. A. Varley	Patented the induction coil
1858	S. A. Varley	First transatlantic telegraph cable
1864	J. C. Maxwell	Equations of electromagnetism

1876	A. G. Bell	Telephone
1879	W. Crookes	Cathode rays and X-rays
1879	E. H. Hall	Hall effect (see glossary)
1880	P. Curie	Piezoelectricity in quartz
1882	T. A. Edison	Pearl St, NY, DC power station
1883	A. E. Dolbear	Radio signals
1887	H. Hertz	Electromagnetic waves
1887	H. Hertz	Antenna (aerial)
1891	J. Trowbridge	Aerial telegraphy by induction
1894	Nikola Tesla	Niagara Falls AC power station supplies Buffalo City
1896	G. Marconi	First wireless link
1897	O. J. Lodge	Inductance for spark telegraphy
1899	G. Marconi	Offers lease of 20 radio sets to US Navy. US asked to buy them. Request refused
1899	G. Marconi	Radio to France across the Channel
1900	J. A. Fleming	Thermionic valve (tube): diode
1901	G. Marconi	Transatlantic wireless link
1902	R. A. Fessenden	Precursor of superhet receiver
1902	A. E. Kennelly	Reflection of radio waves from ionosphere
1902	US Navy	Announced: receivers and transmitters need not be from same maker
1902	Kaiser Wilhelm 2	Marconi's mastery spurs proposal for international radio conference
1904	US Navy	Began daily broadcasts of time signals to help navigation
1906	L. de Forest	Thermionic valve (tube): triode
1906	R. A. Fessenden	Successful AM broadcasting
1906	USS *Chicago*	Sent reliable messages about San Francisco earthquake to the world.
1906	J. G. Pickard	Discovered silicon as a rectifier
1908	H. K. Onnes	Liquid helium, −269°C
1908	Hugo Gernsback	*Modern Electrics*, first radio journal started in USA
1909	Hugo Gernsback	Wireless Association of America (first national amateur radio body)
1911	O. Heaviside	Reflection of radio waves from ionosphere
1911	H. K. Onnes	Superconductivity at −269°C
1912	E. H. Armstrong	Feedback: amplification × 1000

1912	David Sarnoff	On 4.4.1912 heard signals from sinking ship *Titanic* and stayed at his post for 72 hours
1917	L. Levy	French patent of superhet
1924	A. W. Hull	Magnetron
1924	E. V. Appleton	Kennelly-Heaviside layer verified
1924	E. V. Appleton	First radar experiments
1924	J. L. Baird	Televised objects in outline
1924	UK Post Office	Ceased issuing licences for spark transmitters
1926	H. J. Round	Thermionic valve (tube): tetrode
1926	H. Yagi	Yagi antenna
1928	B. D. H. Tellegen and G. Holst	Thermionic valve: pentode
1928	J. L. Baird	Demonstrated colour TV
1929	J. L. Baird	German post office's TV licence
1929	Hans Berger	Records first human EEGs at Jena
1933	E. H. Armstrong	FM: early hi-fi
1934	H. Fletcher	Stereo sound in Bell's labs.
1935	O. Heil	Unipolar transistor (FET) conceived
1935	H. S. Burr	Publishes *Electrodynamic Theory of Life*
1937	Marconi EMI	TV system adopted by BBC
1939	J. T. Randal and H. A. Boot	Magnetron
1939	S. D. Kirlian	'Body-halo' photography in USSR
1943	P. Eisler	Printed circuit patented
1947	Dennis Gabor	Invented holography
1948	J. Bardeen and W. H. Brattain	Point-contact transistor
1950	W. B. Schockley	Junction transistor
1952	W. B. Schockley	FET (field effect transistor)
1955	W. B. Schockley	Electromagnetic imaging scanners
1955	W. B. Schockley	MIC (Microwave integrated circuit)
1957	Herman Schwan	US 'safe' limit of 10 mW/cm^2
1957	USSR	Launch of Sputnik 1
1958	J. Kilby and R. Noyce	Monolithic integrated circuit
1959	J. Hoerni	Planar transistor
1960–4	J. Kilby and R. Noyce	Logic circuits
1961	Robert Becker	Effects of EM fields on growth
1964	Texas Instruments	Patented MMIC (monolithic microwave integrated circuit)

1964	David Sarnoff	LabsCMOS technology
1971	INTEL	4004 chip, first microprocessor
1974	Hewlett Packard	GAs integrated circuits
1975	R. P. Blakemore	Bacteria in water swim towards the magnetic pole
1976	Egon Eckert	Sudden infant deaths because of EM fields?
1976	Plessey	First MMIC built
1976	Dr Stephen Perry	Study of suicides and EM fields
1977	Paul Brodeur	Publishes *Zapping of America*
1978		First videotex sets in UK
1978		Bioelectromagnetics Society founded in USA
1980	Thomson-CSF and	First HEMTs (high electron mobility transistors)
1983	J. B. Fisk Fujitsu	Developed microwave magnetrons at Bell Telephone Labs
1985	Dr R. Becker and G. Selden	Publish *The Body Electric*
1986	J. G. Bednorz and K. A. Müller	Superconducting ceramics
1987	K. P. Chu and Maw-Kuen Wu	Superconducting ceramic, −176°C
1988	Reiter *et al.*	EM fields inhibit melatonin in unborn rats
1989	Paul Brodeur	Publishes *Currents of Death*
1989	Cyril Smith and N. Best	Publish *Electromagnetic Man*
1990	R. Coghill	Publishes *Electropollution*
1990	R. Coghil	Discovery of buckminsterfullerene (C60 fullerene), aiding superconductivity
1993	Paul Brodeur	Publishes *The Great Powerline Coverup*
1993	Studholme family	Sued Norweb (UK) for death of son from leukaemia

Appendix B
Selection of electromagnetic pollution sources

Natural: on Earth

Rain, snow, hail, etc.
Storms, magnetic or lightning
Thunderstorms worldwide (static)

Natural: from outer space

Cosmic noise
Radio stars
Sun, disturbed or quiet

Man-made sources

Electric power

Dirty insulators
Faulty transformers
Faulty grounding
Generators: motors
Step-up or step-down transformers
Power lines acting as antennas

Communications

Aircraft
Amateur radio or TV: CB radio, taxis
Broadcasting, AM, FM, etc.
Cellular radio telephone
Facsimile
Ionospheric scatter
Land mobile radio
Maritime radio
Microwave relays
Personal communication services (PCS)
Radio control of toys
Satellite relays
Telegraphy
Telemetry
Telephony
Tropospheric scatter
Wireless local area networks

Radio help to navigation

Beacons
Instrument landing systems
Loran, Omega, etc.

Radar

Air searching
Air traffic control
Harbours
Mapping
Police speed control
Vehicle distance measurement
Weather monitoring

Tools and appliances

Air conditioners
Blenders
Cookers, microwave ovens
Deep freezers or fridges
Fans
Food mixers
Hair dryers
Lawn mowers
Pumps
Sewing machines
Vacuum cleaners

Office machines

Adding machines
Card punches: card readers
Cash registers
Calculators
Computers
Reprography, photocopiers
Typewriters

Industrial machines

Cranes
Fork-lift trucks
Lathes
Lifts
Milling machines
Printing presses
Presses
Screwcutting machines
Travelators

Power tools

Band saws
Circular saws
Dielectric heaters
Drills
Grinders
Heaters, gluers
Induction heaters
Industrial computer control
Machine or process controllers
Plastics preheaters or welders
RF stabilized welders
Sanders
Teletype machines
Welders and heaters

Vehicles

Aircraft
Alternators
Cars: lorries
Engines
Farm machines: harvesters, etc.
Garage door openers
Generators: motors
Ignition systems
Motorbikes
Outboard motors
Lawn mowers
Portable saws

Lights

Computer displays
Display signs
Faulty incandescent lamps
Fluorescent tubes
Light dimmers
Neon lights

Medical equipment

CAT (computerized axial tomography)
Defibrillators
Diathermy
MRI (magnetic resonance imaging)
X-ray machines
Ultrasonic equipment, including cleaning

Index